Workplace Involvement in Technological Innovation in the European Community

Volume I:

Roads to Participation

EF/93/03/EN

Title Series

Volume I   Roads to Participation
           ISBN 92-826-5669-1   Cat. No. SY-17-93-001-EN-C
           ECU 31,50
           (Revised report originally produced as research report
           Cat. No. SY-72-91-035-EN-C   ISBN 92-826-3186-9

Volume II Issues of Participation
           ISBN 92-826-5672-1   Cat. No. SY-17-93-002-EN-C
           ECU 31,50

Workplace Involvement in Technological Innovation in the European Community

Volume I:

Roads to Participation

Dieter Fröhlich
Colin Gill
Hubert Krieger

European Foundation for the Improvement
of Living and Working Conditions
Loughlinstown House, Shankill, Co. Dublin Ireland.
Tel: +353 1 282 6888 Fax: +353 1 282 6456 Telex: 30726 EURF EI

Cataloguing data can be found at the end of this publication

Revised report originally produced as research report
Cat. No. SY-72-91-035-EN-C   ISBN 92-826-3186-9

Luxembourg: Office for Official Publications of the European Communities, 1993

ISBN 92-826-5669-1

© European Foundation for the Improvement of Living and Working Conditions, 1993.

For rights of translation or reproduction, applications should be made to the Director, European Foundation for the Improvement of Living and Working Conditions, Loughlinstown House, Shankill, Co. Dublin, Ireland.

Printed in Ireland

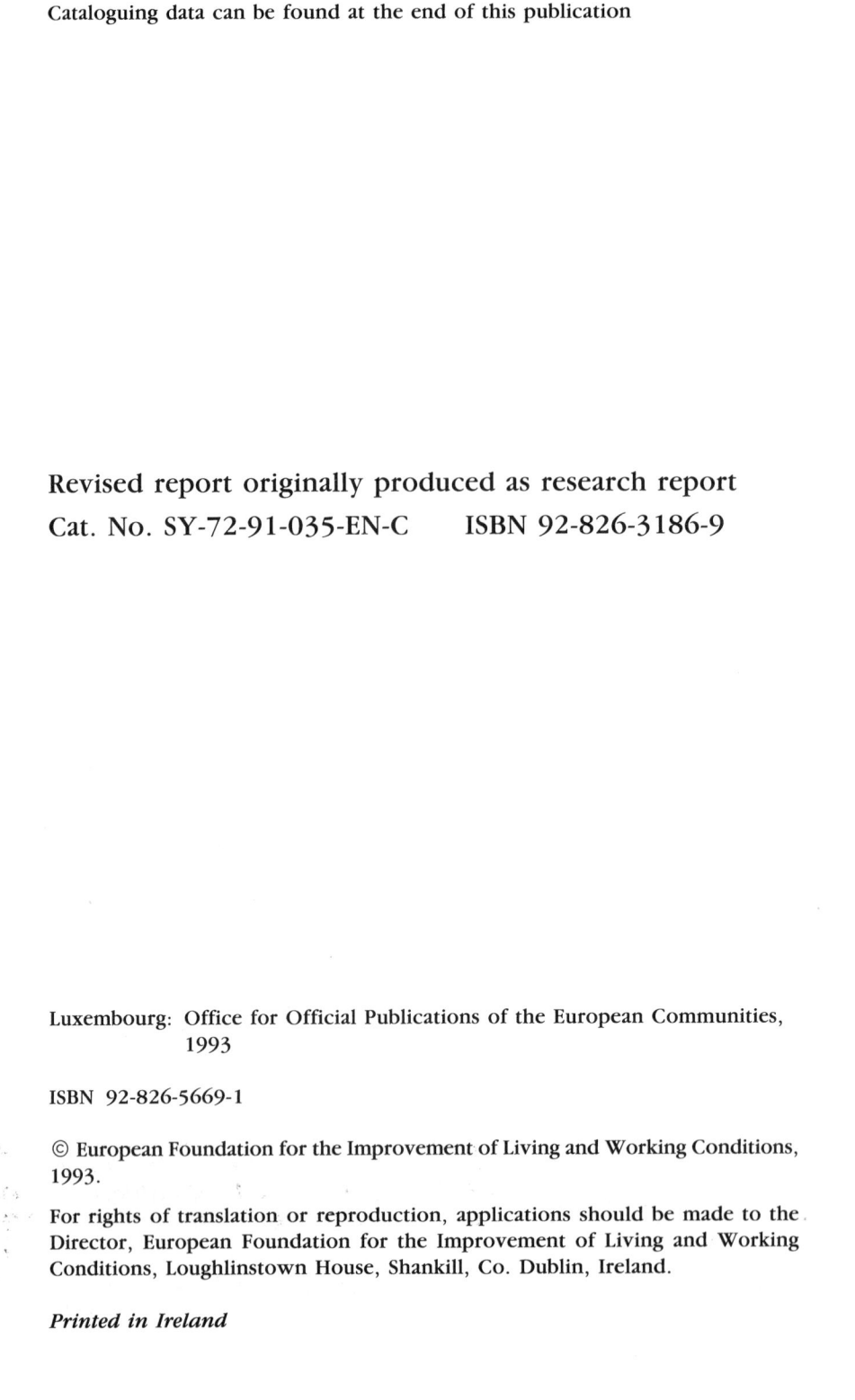

Preface

The 1980s have seen a widespread debate on workplace involvement in the process of technological innovation in Europe. The two volumes of this report present the results of an attitudinal survey on the impact of new information technology in enterprises across the 12 Member States of the European Community and outline and compare the experience and opinions of nearly 7,300 managers and employee representatives.

The introduction of new information technology in European enterprises during the 1980s has had profound implications both for employees and the way such enterprises have been managed. Participation in technological change has been accorded a high priority by the Social Partners and thus the survey gives an unprecedented opportunity not only to evaluate the effects of these changes stemming from new technology, but also to give some detailed information on the degree of participation by employee representatives in such technological change.

The **first volume** of the report "**Roads to participation**" is focused on participation during the different phases of the process of technological innovation. The introduction of new technology can be viewed as going through a number of phases: planning, selection, implementation and post-evaluation. The first two phases are strategic, the latter two are operational.

The **second volume** of the report "**Issues of participation**" produces valuable results on the intensity of participation on different issues directly related to technological innovation. In particular, the survey covers participation in training, health and safety aspects of new information technology, work organisation, investment, product quality and the benefits of participation to both managers and employee representatives. It also gives an overview about the perceptions of the respondents regarding the impact of new information technology on employment levels, particular occupations, employment flows within enterprises, the problems that new information technology poses for particular occupational categories, and the role of participation in dealing with these problems.

Perhaps the most important conclusion that emerges from the survey is the sheer diversity in the forms of participation that exists in the individual countries. Taking all explanatory factors together, it is clear that the types of participation that exist in Europe, or which will emerge in the future, are a product of the way that each Member State's industrial relations system has been shaped by wider political, economic, social and historical forces.

European Foundation for the Improvement of Living and Working Conditions	**European Foundation for the Improvement of Living and Working Conditions**
Clive Purkiss	Eric Verborgh
Director	*Deputy Director*

Table of Contents

SUMMARY 1

CHAPTER I—INTRODUCTION 11
1. The challenge of New Information Technology 12
2. New technology and participation:
   The position of the political actors 14
3. The social dialogue within the European Community 16
4. Major research questions of the survey 17
   4.1 European Trends 18
       (a) Actual extent of involvement 18
       (b) Quality and channels of information 18
       (c) Impact and evaluation of involvement 19
       (d) Potential for increased involvement 19
   4.2 Country specific analysis 20
       (a) Past involvement: 20
       (b) Potential for increased involvement 20
       (c) Factors influencing country specific differences 21

CHAPTER II—NEW INFORMATION TECHNOLOGY AND ITS POTENTIAL FOR PARTICIPATION 23
1. Some characteristics of new information technology 24
2. Social factors affecting technology application 26
3. New information technology: Problems for the parties involved 28
   3.1 Management problems 29
   3.2 Union problems 30
4. Information technology and participation:
   The strategies of the parties involved 31

CHAPTER III—SURVEY METHODOLOGY 37
1. Countries and Sectors 38
2. Survey Procedure 39
3. Limitations of the study:
   industrial climate and sectoral distortions 40
4. Limitations of the study: Screening procedure 41

CHAPTER IV—PAST PARTICIPATION AT THE EUROPEAN LEVEL 43
1. Introduction 44

2. Extent and timing of past involvement 47
3. The content of past participation 52
4. The information issue 54
5. Conclusions 60

CHAPTER V—THE PERCEIVED EFFECTS OF PARTICIPATION 63
1. An overview of the costs and benefits of participation 65
   1.1 Effects of involvement on decision-making 67
   1.2 Industrial relations and mutual understanding in the firm 67
   1.3 Workforce identification and the concept of company culture 69
2. Effects of past involvement dependent on intensity and timing of involvement 72
3. Conclusions 81

CHAPTER VI—FUTURE PARTICIPATION 85
1. Introduction 86
2. Intensity and timing of future participation—the aggregate level 88
3. The dynamics of participation—the individual level of comparison 93
4. Future participation in management and worforce concerns ... 100
5. Conclusions 104

CHAPTER VII—COUNTRY SPECIFIC ANALYSIS 107
1. Introduction 108
2. Country specific factors affecting involvement in the introduction of new technology 108
   (a) Management's reliance on its workforce to achieve its objectives 109
   (b) Management style and attitude towards participation 109
   (c) The bargaining power of organised labour 109
   (d) Regulation 109
   (e) The degree of centralisation of the industrial relations system 109

CHAPTER VIII—A COMPARISON OF PAST AND FUTURE PARTICIPATION IN TECHNOLOGICAL CHANGE IN THE 12 MEMBER STATES 117
1. DENMARK 128
2. GERMANY 133
3. IRELAND 142
4. NETHERLANDS 149
5. BELGIUM 156
6. U.K. 162

7. FRANCE 168
8. SPAIN 176
9. GREECE 184
10. ITALY 190
11. LUXEMBOURG 199
12. PORTUGAL 202

CHAPTER IX—SUMMARY OF INDIVIDUAL COUNTRY RESULTS 211
1. Introduction 212
2. A comparative evaluation of the levels of participation in the Member States of the European Community 212
3. Using our explanatory factors to account for differences in participation levels 216
4. Assessing the potential for participation in the future 220
5. The importance of statutory rights and legal regulations 224
6. Assessing the results in the light of the Val Duchesse Joint Opinion of March 1987 226
7. The policies of the ETUC 230
8. The policies of UNICE 232
9. Conclusions 232

TECHNICAL APPENDICES 237

Summary

In order to remain competitive in world markets, business enterprises in the European Community must take full advantage of the potential for increased productivity that is provided by new information technology. To achieve this, a skilled, co-operative and motivated workforce is required and the traditional hierarchies and forms of control that are found in many enterprises need to be revised. More participation by employees and their representatives in introducing and applying new technology can help reap the full benefits of modernisation.

The European Trade Union Congress (ETUC) and European employers' organisation, UNICE, have formulated policies on the issue of participation in technological change. The ETUC aims at a model of industrial democracy, with rights to participation established through legislation. UNICE is open to all forms of participation by employees, but says that these must agree with the traditions and legal practices of the different Member States. The European Commission has adopted the role of a facilitator, rather than an initiator, of participation.

At Val Duchesse in March 1987, the employers and the unions reached an understanding that employees and their representatives should be kept well informed and be consulted during the introduction of new technology.

Against this background, the European Foundation for the Improvement of Living and Working Conditions in Dublin carried out an attitudinal survey on participation in all the Member States. Its objective was to estimate the extent, impact and value of participation in actual situations and to judge the potential for increased participation in the future.

The main premise of the survey was that participation by employees and their representatives in the introduction of new technology could contribute to the smooth functioning of this technology, to improvements in working conditions and industrial relations and to a more balanced distribution of the costs and benefits of modernisation.

The survey was designed to gather information and opinions from managers and employee representatives. It concentrated on mechanical engineering and electronics, to represent the industrial sector, and on banking, insurance, and retailing, to represent the service sector. The interpretation of the results must take account of these sectoral limitations.

The survey was conducted between February 1987 and October 1988 in two waves. There were 7,326 respondents, divided equally between managers and workforce representatives. It is the largest survey so far on employee participation in the introduction of new technology in Europe. The data sampling and presentation was organised by Harris Research (London) and GFK (Nürnberg).

The study adopted a broad definition of involvement in order to take account of the differences between each country. Involvement in the introduction of technology was defined as:

> any participatory procedure or practice, ranging from disclosure of information to consultation, negotiation and joint decision-making, which formally or informally involves the parties concerned with the information of new technology in the discussion of decisions concerning the process of change.

1. Participation at different phases

The introduction of new technology can be viewed as going through a number of phases:

- planning
- selecting
- implementation
- post-evaluation

The first two phases are strategic, the latter two are operational. Data analysis concentrates on the strategic planning phase and on the operational implementation phase.

In the *planning* phase, the involvement of employee representatives was very low. According to two out of five managers there was either no involvement at all or involvement restricted to the mere provision of information; around 10% stated that employee involvement was in the form of consultation or negotiations/joint decisions. Employee representatives gave a broadly similar response.

In the *operational* phase, the pattern was different. 'No involvement' decreased sharply, information provision remained static at around 40%, and consultation occurred in about 20% of the companies. The percentage of respondents who indicated negotiation/co-determination practices was 16-18%, compared with 10% in the planning phase.

There is a notably high degree of consensus between managers and representatives on the different intensities of participation in both phases. In important managerial areas such as defining market strategies and investment criteria, most management respondents (about 60%) agreed that, at the time of the survey, there was no participation. Between a quarter and a third indicated that participation was characterised by information provision. Consultation, negotiation and joint decision-making played only a minor role.

The situation differs in areas traditionally of concern to the workforce, from work organisation to health and safety measures. About half of the management respondents reported that consultation and co-determination procedures were undertaken here.

The process of participation depends on the relevance of the information that employees receive, and the provision of information was the procedure most often cited. Half of all employee representatives received information that was regarded as complete, timely and useful; 13% reported receiving very poor information on all three counts.

The most general result of indirect participation on various issues is that employee representative involvement either had no impact at all or it had positive effects; negative repercussions were rarely reported.

As far as managers are concerned, participation has generally had favourable effects. It rarely slows down the decision-making process; industrial relations are improved; it assists in the quality of decision-making; it enhances the identification of employees with the goals of the enterprise; it pays adequate attention to the concerns of both sides; and it facilitates the utilisation of skills and engenders a greater degree of acceptance of new technology in the minds of employees. The responses of employee representatives produced a broadly similar picture.

In Europe there is overall a positive attitude to the idea of increasing participation in the future. More than half of all managers who did not involve representatives in the planning phase intend to introduce some form of participation in the future, and although the percentage of managers who favour information as the best form of participation remains generally static, there is a significant increase in the number of managers favouring consultation and negotiation/joint decision-making in the future.

At the implementation stage, only a small number of managers don't want involvement of employee representatives, while about 30% of managers favour information-provision, consultation and negotiation/joint decision-making.

Employee representatives have more far-reaching aspirations and non-involvement in both phases is clearly a minority position. Information provision is regarded as inadequate and employee representatives are clearly in favour of consultation and, particularly, negotiation/joint decision-making.

2. The political context

The ETUC and the UNICE have different views about participation. The ETUC favours a model in which the right to co-determination at all levels and in all company matters is legally binding. The UNICE is opposed to such a policy and favours an open situation in which entrepreneurs and managers can practice any form of participation according to existing prescriptions and local traditions. UNICE also rejects the idea of EC regulations on participation applying to all Member States.

Both sides have reached a compromise in the Val Duchesse Social Dialogue—they have agreed that information and consultation should become standard practice in all European firms.

The survey reveals that participation in many European companies falls well short of this standard; this is particularly true in strategic matters, such as the planning of technical change, and in the areas traditionally seen as the prerogative of management, such as investment and market planning. Although consultation and even negotiation/joint decision-making sometimes occur on operational issues, there are still considerable gaps between reality and the standards laid down at Val Duchesse.

This situation, however, is likely to change. To some degree, management's intentions on the involvement of employee representatives have moved away from the practices of the past. This is particularly evident in operational matters, where we see quite a distinct trend towards meeting the standards set at Val Duchesse. There is also a change of attitude to strategic issues, but its extent is rather limited.

In both policy arenas there is positive change, but it is unlikely to result in reaching the standards set at Val Duchesse in many European firms, particularly in strategic issues.

The results also indicate that the trade union model of industrial democracy finds little favour among managements in Europe. Even the majority of employees—many of them union representatives at company level—are hesitant about seeking far-reaching participation rights in the future. And only a small minority wants co-determination in those areas that are traditionally the prerogative of management. Employee representatives generally opt for involvement that goes well beyond what management is prepared to concede but which falls short of aspiring to radical forms of industrial democracy.

At the overall European level, relations between management and employee representatives in high-tech companies have largely been consensual and co-operative, and our survey leads us to suspect that, despite the differences between the two sides, this spirit of consensus is likely to continue.

3. Country differences

The extensive study of the past decade on the implications of technological change for work shows that there are a number of national factors that shape the opportunities for participation in the introduction of new technology. They are inter-related to some degree and strong forms of particular factors may have a significant impact on the level of participation.

There are five principal factors that can help or hinder the opportunities for participation:

1. management's dependence on the skills and co-operation of its workforce to achieve its objectives for introducing new technology;
2. management style and its attitude to participation;
3. the bargaining power of organised labour to force management to negotiate or consult with its representatives in the absence of any voluntary disposition to do so;
4. regulations which lay down participation rights for employees or their representatives;
5. the degree of centralisation of the industrial relations system that exists in the country concerned.

The top-ranking countries: Favourable factors exist in both **Denmark** and **Germany** that assist employee participation; in both countries there are long traditions of cooperative practices between management and trade unions, underpinned by a system of tripartite consultation and state support for jointly agreed goals. Trade unions are accepted by management as partners in the planning of technological change and there is a stock of necessary goodwill between both sides of industry. Management in both countries uses new technology to increase competitiveness, to enhance product quality and improve customer service; in such cases we would expect a high degree of dependence by management of the co-operation of the workforce in problem-solving etc. In addition, there are extensive provisions in both countries for training and retraining in the use of new technology, coupled with effective means for re-deployment and time off for employees to perform their duties, and there is an emphasis on job satisfaction and adequate job design.

The middle-ranking countries: Just below **Denmark** and **Germany** there is a group of three middle-ranking countries: **Ireland,** the **Netherlands,** and **Belgium.**

In **Ireland**, the industrial relations system has become increasingly centralised over the past two decades and the Irish trade union movement has a fairly high degree of bargaining power. On the other hand, there is very little legislation promoting participation and Irish managers generally do not favour participation and have a low reliance on the skills and co-operation of the workforce.

In both the **Netherlands** and **Belgium** there is extensive consultation at national level before any employment legislation is enacted, and both countries have well-established works council systems, with extensive rights given to their members.

The low-ranking countries: As we move down the table, the factors become more and more unfavourable. In descending order, these countries are the **United Kingdom, France, Spain, Greece, Italy, Luxembourg** and **Portugal. If we consider each factor in turn we find that only in France** and **United Kingdom** is there a dependence by management on the skills and problem-solving abilities of the labour force; this explains the comparatively high level of consultation in Britain at the implementation stage.

Management style in all these low-ranking countries is generally unfavourable. In the **United Kingdom** there is evidence of a shift in the past decade away from negotiation towards consultation and management is becoming increasingly paternalistic in its style. However, this serves only to increase the opportunities for employee representatives to be involved at the implementation stage of technological change.

In **France**, the Auroux reforms are working very slowly and management still jealously guards its managerial prerogatives. In countries such as **Spain, Greece** and **Portugal**, there is a legacy of a dictatorial past and management in these three countries is opposed to the legislation on participation which has been introduced by the new democratic governments, while the trade unions are weak, inexperienced and with poor resources.

Apart from the **United Kingdom** and **Greece**, all these low-ranking countries have trade unions that are divided on political or religious lines, and even where there have been recent efforts to forge closer links and promote unity (for example, in **Spain**, **Italy** and, to a lesser degree, in **France**), these have had very little impact as yet on participation. In **Greece**, there has been a recent split in the GSEE.

Although all these low-ranking countries (with the exception of the **United Kingdom** and **Italy**) have various forms of legislative provisions for setting up works councils, in many of them the trade unions eschew participation and prefer to have technological change subsumed under collective bargaining.

The survey results for **Portugal** stand out against those for all other EC countries. This country has the highest level of 'no involvement' in the planning stage and the lowest level of 'no involvement' in the implementation stage. These unusual results are largely due to the very detailed and precise legislative provisions which define the rights for information and consultation for works councils and exclude major issues associated with technological change.

Perhaps the most important result of our survey is that in every country both sides indicated that they wanted more participation in the future. This means, firstly, that the level of 'no involvement' had decreased substantially in the preferences of both sides for the future compared with what existed at the time of our interviews in 1987-88, and, secondly that both sides envisaged more negotiation and more co-determination in both phases of technological change.

There was a substantial shift in opinion among employee representatives—much more so than among managers. In two countries, **Denmark** and **Germany**, there was a marked preference for major decisions on technological change to be taken either through negotiation or joint decision-making. **Portugal** again stands out as having the least shift in opinion.

It is not surprising that the preferences of employee representatives for the future envisaged much higher levels of participation than did those of managers; this was true in all countries. What is perhaps most notable is that this difference is particularly evident in four countries—**Denmark, Germany, Italy** and **Greece.**

In **Italy**, more than three out of five managers envisaged either no involvement at all or simply the mere provision of information, whereas employee representatives had much greater aspirations.

Compliance with the Val Duchesse guidelines: In comparing the results of the survey for each country, we used the benchmark established by the Val Duchesse Social Dialogue—which states that employee representatives must, at the minimum, be consulted about new technology at the planning and implementation stages. Of course, participation that encompasses negotiation or joint decision-making is superior to that laid down in the Val Duchesse provisions.

If we consider the views of managers on past involvement, we find that only in **Denmark** and **Ireland** did half of the managers report that consultation or stronger types of involvement occurred at the planning stage.

A middle-ranking group of countries—which included **Germany**, the **Netherlands, Belgium**, the **United Kingdom, France, Spain** and **Greece**—had no more than 30% of managers who claimed levels of participation that were within Val Duchesse guidelines. There was almost no compliance with Val Duchesse in the cases of **Italy, Luxembourg** and **Portugal**.

As we would expect, the compliance with the Val Duchesse guidelines is better at the implementation stage. Yet only in **Denmark, Ireland** and the **United Kingdom** do we find a majority of managers who state that the implementation stage of technological change is characterised by at least consultation or higher levels of involvement. There is a middle-ranking group of countries where the percentage of managers reporting compliance with the guidelines clusters around 40%. Below that, the percentage decreases to just above 5% (in the case of **Portugal**).

Three immediate conclusions about the future likelihood of the Val Duchesse guidelines being observed can be derived from our data.

Firstly, the most striking conclusion is that in all the countries surveyed there is a significant shift of opinion on both sides in favour of more participation.

Secondly, it is evident that employee representatives would prefer a higher level of participation than their management counterparts; this is only to be expected.

Thirdly, the pattern of preferences for the futures, in terms of the ranking of countries, bears a close resemblance to that of the past.

There are five countries where more than half of the managers indicated that the Val Duchesse guidelines would be most likely observed at the planning stage: **Denmark, Ireland,** the **Netherlands**, the **United Kingdom** and **Greece**. A striking observation that can be made is that **Germany** is not included in this list; only two out of five managers in this country anticipate that the Val Duchesse guidelines will operate in the future.

At the implementation stage, the situation is much healthier. Here there is a similar pattern to that of the planning stage insofar as the preferences of managers are concerned. **Denmark** and **Germany** show the most preference for strong forms of participation. A majority of managers in all countries, except **Italy, Luxembourg** and **Portugal**, envisaged participation that included consultation or stronger forms. On the negative side, the levels of 'no involvement' are still worrying. **France, Greece, Italy, Luxembourg** and **Spain** all have results showing that more than 10% of managers intend to introduce new technology into their enterprises without any employee involvement whatsoever.

Conclusion

Perhaps the most important conclusion that emerges from our survey is the sheer diversity in the forms of participation that exists in the individual countries. Taking all our explanatory factors together, it is clear that the types of participation that exist in Europe, and which will emerge in the future, are a product of the way that each Member State's industrial relations system has been shaped by wider political, economic, social and historical forces.

There is evidence that all countries of the Community are moving in different degrees towards greater participation. Our survey shows that the experience of participation by both sides engenders favourable attitudes towards more intense forms of participation in the introduction of new technology.

CHAPTER ONE

Introduction

This report presents the results of an attitudinal survey on the topic of workers' participation in the process of introducing new information technology in European enterprises. It covers the experience and opinions of nearly 7,300 managers and workforce representatives in all twelve Member States of the European Community. A central focus of this report is how workers' participation in the introduction of new technology is related to industrial relations in the European Community. In this respect, the political actors at Community level (Commission, European Parliament, ETUC, UNICE) and the representatives of the two sides of industry on the company level (management, workers' representatives) have their own specific interests and policies on the implications of the technological modernisation of Europe. This introduction gives an overview of the problems, political positions and policies, and the major research questions addressed in the survey.

1. The challenge of New Information Technology

New information technology based on microelectronics is a world wide phenomenon. In the business world we already speak of the "global village" denoting the fact that new information technology has made instant and continuing communication around the world a reality. The functioning of world markets has been fostered through this technology. The easier access to world markets has not only opened up new opportunities, but also increased competition at the same time. Europe's ability to remain competitive on these markets involves the constant development and implementation of new information technologies.

Technological innovations have always been a feature of social change. The new information technology, however, has distinct characteristics for which there are only a few historical precedents. It is not a technology which is restricted to one sector of industry or one occupation, but can be described as a 'heartland' technology which pervades the world of work and society generally. Nor is it confined to certain sectors of the economy; in short, it is a **pervasive technology**. It has major repercussions for the world of business and it will also change the living and working conditions of people everywhere. The new information technology is seen not only as opening up new opportunities, but also producing considerable anxiety about its effects.

New information technologies appeared in the 1970s at a time of structural crisis of the Western economies with low growth-rates and high unemployment. At first, they were seen as promoting labour saving through further automation of production processes. This new phase of technological innovation generated anxieties which were derived from the knowledge about the impact of automation in the past: job reduction, skill loss, deterioration of working conditions and increased control of management over the workforce.

It is certainly true that new information technology has the *potential* for such a negative outcome. However, it does not necessarily follow that such a pessimistic scenario will become a reality. Indeed, current wisdom on the effects of new technology gives a more balanced view of the likely effects of these technologies. Until about two decades ago many believed that technology determined the organisation of work: there would be only one way of applying a given technology. This **deterministic approach** suggested that within a given technology, specific results must inevitably follow: a certain organisation of production processes, particular job und skill definitions, control needs and hierarchies, and a certain relation of functions which make up the organisation of companies. People, their needs and capabilites had to be shaped to the possibilities and restrictions dictated by technology.

Research findings in the 1970s and 1980s challenged this view of technological determinism and emphasized that there have always been choices for different technological applications. Whilst these choices are limited, there is much more scope for applying technological solutions in different settings. Technology has lost its sacrosanct status of determining organisation structures, skills and skill use, hierarchies etc.

The advent of new information technology has led to a resurgence in the debate about the role of technology in the world of work. It can be used as a labour saving device or as a job creator. It requires new skills and new commitments. It creates uncertainties in organisations as old schemes often prove to be inadequate. It can hardly be imposed but needs careful adaptation to existing institutions in a "process that mediates between differences of interests" of the representatives of the two sides of industry at company level[1]. It often functions only where there is a climate of high trust between management and employees. Thus, new information technology must be seen in terms of its social impact.

The wealth of our Western market economies is based upon the performance of enterprises. It is at this level where management and employees work together which ultimately determines the performance of the entire socio-economic system. Decisions on this level on how to relate technology and the people who work with it and how production and services are organised, mould the outcome of societal development. New information technology means change. Traditional concepts of organisation become obsolete and must be altered. Old schemes of control often have to yield to new relationships of trust and mutual dependence. In this process of social change at enterprise level, industrial relations are also affected. The workforces and their representations may gain a new, more active role.

2. New technology and participation: The position of the political actors

In the European Community the main political actors have formulated their different policies on workforce participation in the process of technological modernisation. Whilst there is a small degree of consent between all parties, there is also a great deal of disagreement.

The **European Trade Union Confederation (ETUC)** has defined its political aims recently in 1985 and 1988. At its fifth congress in Milan in 1985, the ETUC has stated that "effective negotiation of technological change often means influencing the change when it occurs and not merely reacting to the changes later. For the trade unions this means negotiating with management before the introduction has been planned. It means reappraising the knowledge and requirements of working parties in the production process and calling in outside experts to demystify the new technologies." At its sixth ordinary congress in Stockholm in 1988, the ETUC defined its goals more precisely. Here, it asked for

(a) "the right of employee representatives to be fully informed, consulted and also to negotiate on all important company matters before decisions are taken;

(b) equal participation by employee representatives in all company decisions of significance to the workforce;

(c) extension of decision making rights at all levels of decision making according to the organisation of companies ... The employee representatives at all plants must accordingly ... have the right to be informed and consulted on company planning, to negotiate and to represent their interest jointly at the European level."

The ETUC position is strongly opposed by the **European Employers Association (UNICE).** UNICE insists on the right of company management to take final decisions. It further rejects any form of statutory regulation at the European level:

"Employers do not oppose practices such as co-determination, let alone information and consultation. They strongly oppose the imposition of such practices on all member states by means of EC level legislation. Such matters must be worked out by the parties most directly concerned, according to local conditions, traditions and legislation. In this way, the solution arrived at will be better, and the parties will be more deeply committed to making them work."

The bodies of the European Community, the **European Commission, Parliament** and **Council** pursue a policy which tries to reconcile the opposing views of both the ETUC and UNICE. Originally, reconciliation between the different positions adopted by the two social partners was addressed in line with traditional forms of Community intervention. Attempts were made to introduce new directives relating to participation, in particular the so called **Fifth Directive** which intended to regulate the structure and the activities of public limited companies, including workers' representation. The **Vredeling Directive** aimed at providing information and consultation rights for employees in multi-national enterprises. The **European Company Statute** would allow national companies in at least two member states to set up the European Public Limited Company. This statute proposes a choice of three different models of workers' participation.[2]

However, these attempts to regulate and extend workforce participation in Europe by institutional and legal forms of intervention have not been successful, at least for the present. At the moment, the enaction of the Fifth Directive is proceeding very slowly; the Vredeling Directive was first diluted, and then frozen by the council until 1989 for numerous and complex reasons. The European Company Statute received very controversial comments by the two sides of industry and by certain governments of the Member States.

In December 1989, the EC summit in Strasbourg agreed by eleven to one votes on a declaration in regard of a **Community Charter of the Fundamental Social Rights of Workers.** It stipulates that "information, consultation and participation for workers must be developed along appropriate lines, taking into account the practices in force in the various Member States... Such information, consultation and participation must be implemented in due time, particularly in the following cases:

- When technological changes which, from the point of view of working conditions and work organisation, have major implications for the work force, are included into undertakings;
- in connection with restructuring operations in undertakings or in cases of mergers having an impact on the employment of workers;
- in cases of collective redundancy procedures;
- when transfrontier workers in particular are affected by employment policies pursued by the undertaking where they are employed."

This social charter was accompanied by an **Action Programme** which spelt out in detail the planned activities of the Commission up to 1992. One new initiative of the action programme in the field of participation is a proposal of the Commission on a draft directive for informing and consulting the employees of enterprises with complex structures, in particular trans-national undertakings: "The Commission, following consultation with the social partners will prepare a draft for a Community Instrument which in substance could follow the following principles:

- Establishment of equivalent systems of workers' representation in all European-scale enterprises.
- General and periodic information should be provided regarding the development of the enterprise as it affects the employment and the interest of workers.
- Information must be provided and consultations should take place before taking any decision liable to have serious consequences for the interests of employees, in particular, closures, transfers, curtailment of activities, substantial changes with regard to organisation, working practices, production methods, long term co-operation with other undertakings, etc.
- The dominant associated undertakings should provide the information necessary to the employers to inform the employee representatives."

The action programme stresses the belief that practices of good information and consultation of workers and their representatives is an explicit potential objective of the Commission and of most members of the Council of Ministers.

3. The social dialogue within the European Community

The use of directives designed to establish general European regulations seem difficult to develop and equally difficult to enforce. In the light of this, new principles of regulation have been developed in the past for which the formula of the **Social Dialogue** is used. The principles of this social dialogue are:

- a progressive approach to agreed solutions rather than the imposition of pre-ordained models;
- the use of flexible instruments like collective agreements, framework agreements and joint declarations rather than rigid solutions like directives;
- the primary initiative should be left to the social partners to search for common ground, with the European Commission playing a role of facilitator rather than of initiator.

Following this new approach, a first conference was held in **Val Duchesse** in November 1985 on the initiative of the Commission, and new avenues for discussion between the social partners were opened. Eventually, in March 1987, these discussions led to the expression of a **Joint Opinion** on information and consultation as to the introduction of new information technologies in firms. In this document the social partners "recognised the need to make use of the economic and social potential offered by technological innovation in order to enhance the competitiveness of European firms and strengthen economic growth thus creating one of the necessary conditions for better employment and, taking particular account of progress in the field of ergonomics, for improved working conditions...".

Both sides take the view that, when technological changes of major consequences for the workforce are introduced in the firm, workers and/or their representatives should be informed and consulted in accordance with the laws, agreements and practices valid in the Community countries. This information and consultation

- "must facilitate and should not impede the introduction of new technology, the final decision being exclusively the responsibility of the employer or of the decision making bodies of the firm. It is understood that this prerogative does not exclude the possibility of negotiation where the parties take a decision to that effect."

Both sides also agreed that information and consultation must be timely:

- "Information means the action of providing the workers and/or their representatives at the level concerned with relevant details of such change, so as to enlighten them as to the firm's choices and the implications for the workforce.
- Consultation of the workers and/or their representatives at the level concerned, means the action of gathering opinions and possible suggestions concerning the implications of such changes for the workforce, more particularly as regards effects on their employment and their working conditions."

Reviewing the political concepts of the actors on the European level, the Joint Opinion of Val Duchesse seems to be a workable approach in the European Community at present. It is distinct as to information and consultation practices in the enterprises and leaves room for manoeuvre for both sides. The openness of the approach permits solutions to the issue of workforce participation in technological change which go beyond the stipulated measures. From a positive perspective, it can be regarded as a first, open concept for social change within European enterprises.

4. Major research questions of the survey

The unprecedented flexibility of new information technology and the need for social change in enterprises have made workers' participation a **priority issue** of industrial relations policy within the European Community. In this context, the **European Foundation for the**

Improvement of Living and Working Conditions in Dublin carried out an attitudinal survey in all member states of the European Community in 1987 and 1988. The survey was designed to measure the opinions of both managers and workforce representatives in selected industries about their perception and experience of participation in the process of introducing new information technology in their enterprises. The **main topics** approached in this survey are the extent and the impact of participation so far. It further aims at finding out the potential for increased participation of employee representatives in the future on the basis of past experience. In the survey, equal weight was given to the responses of both sides.

The overall structure of the report can be divided in **two main parts**: the first focussing on the overall Europe results, and the second part analyses the differences and areas of agreement in each of the twelve Member States of the Community. Four central aspects of the involvement of workforce representatives in the process of technology introduction are covered:

(a) extent, timing and content of involvement in the past;

(b) quality and channels of information given by management;

(c) impact and evaluation of involvement by the parties;

(d) potential for increased involvement in the future.

4.1 European Trends

At the European level, the following issues and questions should be answered according to the data generated by the attitudinal survey:

(a) Actual extent of involvement:

- What was the actual extent of workforce representative involvement in the European Community in 1987-1988? What is the extent of participation in the different phases of technological innovation, particularly in the strategic phase of technology planning?

- How do the two parties perceive the intensity of past involvement? Do both sides have different or common views on the actual level of participation?

- How does the actual intensity of involvement relate to the joint opinion reached at Val Duchesse? What is the proportion of companies which provide information and consultation in the process of technological innovation in Europe?

(b) Quality and channels of information

- How do employee representatives evaluate the information provided by management?

- Does the quality of information enable employee representatives to really grasp the situation and contribute to solutions?
- To what extent does the usefulness, the amount and the timing of information provisions fulfil the expectations of employee representatives?
- Does the quality of information supplied by management fulfil the criteria which have been set by the two sides in the joint opinion of Val Duchesse?
- Does the provision of information flow through normal, established channels or are new channels established, for example, by technology agreements for information on technical change within the enterprises?

(c) Impact and evaluation of involvement

- Do both parties perceive decision making to be influenced by participation procedures? Does participation retard or accelerate decision processes?
- Does participation have a positive impact on productivity and flexibility of production? Does participation retard or accelerate the implementation of new information technologies in enterprises?
- What is the impact of participation on the structure of power and of the system of industrial relations in the company? Does participation provide a base for better mutual understanding between the two sides of industry? Does participation provide a base for a better mutual understanding or is it perceived to only be a means to institutionalise change in the power balance within companies which might lead to further conflict in the future?
- Does participation improve the acceptance of new technology by the workforce and thereby facilitate its implementation and smooth functioning?

(d) Potential for increased involvement

- How much future involvement of workforce representatives do both sides envisage?
- How far did both sides change their attitude towards participation in the future under the impact of participation experiences in the past?
- To what extent can we observe different strategies of the practitioners at the company level (personnel managers and employee representatives) from their national and international representative bodies such as ETUC and UNICE?

- To what extent is the ETUC concept of early, design oriented and highly intensive involvement of a bargaining and co-determination type in all phases of technological change supported by workers' representatives on company level in the European countries?
- To what extent do management and employee representatives support the position of the European Commission for effective information and consultation practices in Europe?
- How does management in the sector of new information technology relate to the UNICE position regarding employee representatives' involvement? To what extent do they envisage complying with participation practices agreed upon in the Joint Opinion of Val Duchesse?

4.2 Country specific analysis

The second main part of the analysis covers some of these issues on the level of country specific analysis. From a political point of view the following questions may be important:

(a) Past involvement:
- To what extent do the companies in the separate countries differ as to the current extent and timing of participation in the phases of technology introduction?
- What are the political implications of the existing country specific differences? Do these differences support the position of the European employers who warn of the negative repercussions of Euro-wide regulations under the circumstances of prevailing national traditions of industrial relations? Or, do they support the position of ETUC and the European Commission which advocates the long-term necessity to harmonise participation practices via Community regulations or voluntary agreements to harmonise the economic and social development of Europe?

(b) Potential for increased involvement
- What intensity of participation in technological innovation does management envisage for the future in each of the twelve Member States? How far will it deviate from the Val Duchesse benchmark and from the political strategy of UNICE? How do the Member States rank in this respect?
- Do we find a division of management strategies in Europe in two extremes: a co-operative, pro-active Scandinavian- German approach enlightened by the ideas of participation as a productive force and a southern European and United Kingdom reactive, conflictual approach which minimises involvement at the level of power relations between the two sides of industry?

- How far do workforce representatives deviate from the approaches of trade unions towards participation in technological change laid down at the congresses in Milan and Stockholm in 1985 and 1988? Do we still find the split between a co-operative Scandinavian-German approach on one hand and a conflictual, non-participative approach of southern countries and in the United Kingdom and Ireland?

(c) Factors influencing country specific differences

- What is the relation between the legal framework of regulations on participation and the actual practice of participation in the Member States? Do legal provisions increase the actual extent of involvement?
- Can we observe any difference in the intensity of participation between centralised and decentralised industrial relations systems?
- What is the impact of tradition within the national systems of industrial relations on the actual practice and the future strategies of the parties involved for participation in technological change?

NOTES

1. OECD: New Technologies in the 1990s. A Socio-economic Strategy. Report of a Group of Experts on the Social Aspects of New Technologies, Paris 1988, pp. 24 - 25.

2. The Commission proposed three alternative formulae for workers' participation in the management of a European company: a German system, in which workers are represented in managing bodies; a Franco-Italian system of enterprise committees where workers'representatives come together, separately from managing bodies; and a Swedish system where each company defines co-management rules through an agreement negotiated with workers' representatives. However, companies would not be able to opt for provisions circumventing stricter national legislation already applying.

Chapter Two

New Information Technology and its Potential for Participation

As we saw earlier, new information technology has been characterised as a pervasive technology with repercussions that are not restricted to the business world but which will extend throughout society. The introduction of these technologies implies social change alongside technical change. To develop the full productivity and profitability potential of the new technologies to produce higher income and better working and living conditions for the workforces, changes of organisational structures and traditional attitudes are necessary. At societal level, long established traditions of business life and industrial relations practices are affected. At the level of the enterprise, managers, employees and workforce representatives are those who have to mould social change. In this chapter we attempt to clarify in more detail the main new features of new information technology and their possible repercussions in the world of work. We also seek to highlight particular problems which affect management and unions, and how both sides intend to cope with them. Finally, we attempt to present a possible strategy to reconcile divergent interests.

1. Some characteristics of new information technology

Compared to previous forms of technological innovation, new information technology is considerably more flexible and offers much more choice in how it is applied. One important characteristic of this technology is its ability to transmit, store, and process information. In the production process, its basic feature is that of data transmission, filing and processing on which the whole organisation of the enterprise is centred. Physical work, such as welding and paint-spraying, is more or less taken over by "intelligent" machines while man's part consists of preparing and programming the robots, watching and regulating the automated process, and tending and repairing the machinery. Production work takes on an indirect character[1]. The advantage of this type of automation in comparison to the older types lies in the flexibility with which the production of new goods can be reprogrammed. Almost the same technical equipment can be used for different products, and the time needed to re-programme and adjust the machinery is comparatively short.

The problem in applying this technology centres around its organisational prerequisites and consequences: **The more complex the technology applied, the greater its repercussions on all divisions of the company.** An isolated single CNC-machine-tool, a single text processor or an electronic cash register is fairly easy to integrate into the whole existing system of production or administration. The difficulties increase with more complex technologies like Flexible Manufacturing Systems (FMS)[2]. This form of production unit is capable of producing a range of discrete products with a minimum of manual intervention. It consists of machine tools, assembly machines and other operative machinery, linked by a shifting system to move components from one work station to another. It works as an integrated system under a programmable-type control, and it is capable of producing families of products in small batches.

To run such a system effectively, it is advisable to design the products in a way that the design parameters conform to the later programming needs of the production process or, even better, generate production programmes along with the design of products. In doing so, one tries to use CAD/CAM techniques[3], a further step in process integration. When applying these sophisticated techniques, the supporting departments of the company are drawn into the logic and prerequisites of the production technologies. **To make full use of the flexibility potential of the new technology, marketing, administration, stocks, etc. must be prevented from becoming production bottlenecks.** Accordingly, these functions, too, tend to be reorganised to meet production needs. Even suppliers of the company may be forced to adapt to the company's technical standards and requirements.

As regards office work, new information technologies are challenging the older type of computer applications which have a tradition most of all in the banking and insurance sectors. This type of office automation followed, in principle, the pattern of industrial automation, using large central computers ("main frames"). To a large degree, these computers shared the inflexibility that was characteristic of production process technology. The current situation is characterised by electronic networks where data are available and can be processed not only on a company level but nationwide and internationally. In contrast to older organisational schemes, data entry and data production do not need to be separate any longer. Employees may carry out both, and this may result in a decentralisation and de-specialisation of tasks, and the worker may assume a more fully-rounded role.

One of the major effects of the new technology is to permeate the whole of the enterprise. As we have seen, it is not simply the case that new technology substitutes human labour with a form of automation, important though that is. It is the **information** part of information technology that is significant. Only if this information is appropriately used, can its flow enable enterprises to adjust to the challenges of increased competitiveness; failure to so do could have catastrophic consequences. Information technology also serves to highlight any inherent weaknesses in an enterprise's structure, as well as its managerial capabilities and its market position. This will be true whether it uses microelectronic-based equipment in its production, administration or service activities.

Given the need for enterprises to make the fullest use of information technology, enterprises need to restructure themselves in many different ways. Information technology frequently requires companies to adapt it to their own requirements, sometimes on a trial-and-error basis. This sometimes means that management has to face problems of considerable complexity and uncertainty.

Uncertainty marks the planning stage of technology introduction which sometimes extends over years. But uncertainty and complexity might persist even under conditions of technology application. As most repetitive-type work is taken over by technology, the remaining work is mainly connected with the phases of preparation, regulation and maintenance. The resulting tasks are no longer the sum of a number of predetermined microactivities, but have a different quality. Work that is carried out in such organisational settings needs a workforce able to exercise independent judgement, personal flexibility and a sense of responsibility for the proper functioning of the work process. These are professional attributes that are not of a technical but of a managerial nature. Here, great importance is attached to participation, which presupposes that workers have a new relationship to the objectives of efficiency, productivity and quality.

2. Social factors affecting technology application

The description of organisational settings and skill and motivation requirements given above mark the positive extreme on a continuum of technology implementation. It indicates the **potential** of information technology for new organisational solutions. This potential can only be realised under certain conditions. If such conditions are lacking, new information technology might as well be organised in a rather restrictive and traditional way. Between the two extreme solutions of technology application, a variety of approaches is feasible, and they depend on different factors:

(a) The **skills of the available workforce** have been shown to influence the way in which information technology is applied. In the case of a rather poorly qualified workforce either in the enterprise or in the labour market, companies seek to find rather simple technical solutions which might produce sub-optimal results but which fit the given situation. Cross-national studies have shown that the application of new technology in the production sector in the USA, the United Kingdom and in France is generally less sophisticated and follows traditional lines of mass production more often than is the case in Germany. This can be explained by the different systems of vocational education which make for a lower level of average skills in the USA, United Kingdom and France than in Germany[4]. Thus, employers and managers often choose a technical solution which is compatible with the average skill level available in their countries thereby perpetuating traditional organisational structures and technical applications.

(b) **The objectives that management has for introducing new technology** will obviously vary from one organisation to another. In most cases management will have several goals in mind when introducing new technology into their organisation. The emphasis between these is likely to vary according to the priorities and purposes of their organisation and the context in which the organisation operates. These goals include increased competitiveness, greater control over the work process, reduced costs, enhancing product quality, improving customer service etc. Such different management goals will have repercussions both on the type of technology applied and the way company organisation evolves as a result.

For example, companies which concentrate their activities either on the sale or production of standardised goods or services, which operate with mature price competitive markets or where office technology is used to reduce costs by standardising the collection or processing of information will utilise process innovations. The effects of such technological change in cases such as these are likely to lead to a more rigid segmentation of work and a greater concentration of programme-related functions which are carried out by specialist departments. In other words, such goals foster traditional patterns of the division of labour. They tend to preserve or even reduce traditional skill requirements and traditional control schemes and hierarchies within the organisation.

Conversely, where management is concerned to enhance the competitiveness of their organisation by offering its customers improved levels of service and enhanced product quality, the new technology introduced will serve to highlight product performance, quality control, design requirements, technical sophistication, more flexible scheduling and greater emphasis on customer requirements. Such management goals require more flexible technical systems and a flexible organisation with adaptable and broadly skilled employees who are adept at problem-solving and whose skills and co-operation are highly valued.

(c) Finally, the organisation which introduces new technology is influenced by the **vested interests of powerful groups of actors in the enterprises**. Each workshop must be regarded as a political arena in which sections of the workforce pursue their own interests. In the process of technological modernisation they will seek to preserve or create organisational solutions which serve their own interests; qualified workers will look to technology applications which allows their skills to be preserved or enhanced and which do not block career developments. Separate specialist departments will seek to implement new technology in a way which permits its members to carry on as specialists. Thus, technology applications can be fostered which preserve an organisation characterised by a division of labour and segmentation of workforces where, in fact, a more flexible organisation with more fully rounded roles for the employees might be more appropriate to achieve optimal productivity results.

Middle management might act as such a group to influence technology application and work organisation. Middle managers are the people who collect, process, and pass information up and down the hierachy and who exercise control. Such functions can best be retained in the scope of traditional, more rigid mass production type of technology application. Where more flexible technology is introduced, a higher commitment of employees is needed. For middle management, this might mean a threat to their functions and their employment. They, too, might therefore be inclined to influence technological innovations in a way that suits their interests best. In such a case it would mean preserving a pyramidal, company hierarchical structure introducing a more "conservative" type of new technology application.

These instances illustrate that new information technology can be applied quite differently, and that social factors like the available skills, management goals and interest of the actors involved all play a role. The potential of new information technologies for flexible organisational solutions creates opportunities for the actors to influence the process of technological innovation and its outcome according to their own needs: to avoid possible hazards which are perceived to be detrimental to their own interests and to arrive at solutions more to their liking. It is this flexibility in technology applications that gives workers, workforce representatives and unions greater opportunities of participation than at present.

3. New information technology: Problems for the parties involved

Technological innovation means change, and introducing a pervasive technology such as new information technology naturally accentuates the scope and depth of such changes. At the level of the enterprise and the workplace, existing organisational structures are likely to become obsolete. At the same time the search for new solutions creates its own new problems and dangers.

3.1 Management problems

It is the function of management to ensure the survival and growth of their organisations. These objectives are attained by keeping their enterprises profitable. The introduction of new technology is essential to preserve and enhance profitability. As we have indicated earlier, diverse factors such as the availability of skills and company goals influence the way these technologies are applied. It can be done more defensively preserving known structures or more progressively to make full use of the flexibility potential of new technology and thereby to reach optimal productivity. Whatever solution is chosen, it means that management has to cope with uncertainty. However, the new technology serves to heighten uncertainty, particularly if the company embarks on fully utilising the flexibility potential of the new technology.

In order to achieve a more flexible organisation management faces a dilemma: it has to **reduce its direct control of the workforce**, and allow the workforce greater discretion to use its own intitiative at work. Managers who are used to traditional schemes of supervision and control often feel threatened by this new situation. The upper echelons of the company hierarchy become more dependent on those below. Such dependency on the workforce is only feasible **if management can rely on the motivation and compliance of workers and employees**.

In many cases, the optimal use of the new technology requires greatly enhanced skills. These skills are not only centred around the handling of the technology itself, but encompass such intangibles as the capacity for abstraction, for working in teams, for communication between levels of managerial hierarchy and between different functions. The problem of **skills, skill acquisition and training** therefore requires urgent attention.

Particular problems might arise during the process of technology introduction as a result of the **industrial relations** climate in the enterprises. As the companies operate in a situation of high uncertainty as to the proper technical solution, including hardware, software, organisational schemes and their repercussions in the future, technological innovation contains elements of a trial-and-error process. Final solutions have to be arrived at step by step, and from a management point of view, **flexibility is needed** to constantly review the appropriateness of particular solutions. The case studies in preparation for this survey have shown that an important precondition for management's cooperation with workforce representatives is the acceptance of both sides that detailed regulations and formal agreements as to the future organisation of work can be counter-productive for both parties. Management is therefore reluctant to commit itself too far in advance to detailed commitments on future forms or work organisation and often fears that the **advantages that new technology offers can sometimes be eroded by a long drawn-out bargaining process**.

3.2 Union problems

Trade union perception is obviously different. They must preserve the interests of their members and limit any threats for the workforce as a consequence of technology introduction. Given the pervasive effects of new information technology on potentially all aspects of company structure, such threats are numerous:

Job loss and **job protection** are at the fore-front of trade union concern about current technological change.

New information technology emerged with the reputation of a "job killer". This reputation did not only stem from the undoubtedly high productivity potential of new technology, but also from the way the technology was promoted by the manufacturers of technological equipment. This threat loomed particularly large as the new technology gained ground in enterprises at a time when Western economies were facing a structural crisis, with high rates of unemployment and a decline in union influence.

At first, new information technology led to apprehensions about **deskilling.** The downgrading of skills was a feature of traditional automation in manufacturing. New information technology was therefore seen as being likely to intensify this process of deskilling. In addition, new technology was being introduced into the services sector and its office work - generally areas which had hitherto been unaffected by automation technologies. The trade unions warned that deskilling and "Tayloristic" forms of work organisation could emerge in these sectors, too.

Coupled with the skill problem is the question of **workforce recomposition** and **the segmentation into core and peripheral workforces**. The printing industry illustrates how a whole strata of occupations are in danger of disappearing. If the new technologies are implemented in a flexible way they need a core group of highly skilled workers to monitor the technology while a smaller number of people carry out tasks which have not yet been automated. Thus, a split is being created in the workforce, between those who remain in the system, enjoying the benefits of technological innovation and adapting successfully to the changed requirements of the job (core work-force), and those who lack that ability and, as a result, are gradually forced out of the productive cycle or who are subject to infrequent employment in less attractive jobs. Women in certain occupations were seen to be particularly affected by the process of change. There was also the perceived threat of 'marginal' workers, such as part-time workers, agency workers, homeworkers, etc[5]. To avoid such splits in workforces, **training** and **re-training** become important issues. Traditionally, these questions did not feature as a high priority for trade unions. Nevertheless, given that only the acquisition of new skills will reduce the danger of employees being pushed towards the peripheral workforce or even out of enterprises, the training problem assumes a heightened importance for unions.

The under-utilisation of skills was a major grievance in many manufacturing establishments in the past. Today, unions face the opposite danger: over-utilisation of skills. Work with new information technology implies a change of emphasis from physical-manipulating tasks to mental-conceptual commitments. The growing need to concentrate and an increasing level of responsibility for the smooth functioning of the production process and for expensive technical equipment result in a **change of job conditions**. It is generally agreed that working with new technology leads to a reduction of physical stress in work at the expense of increased mental efforts and more mental fatigue. New health hazards through working constantly with visual display units (VDUs), for instance, add to the list of problems.

The issue of **pay levels** is another unsettled question. The competent operation of new technology normally necessitates additional training and the assumption of new responsibilities. These new skills and responsibilities frequently require pay systems to be radically altered. As discreet manual work decreases and workers are more likely to be monitoring and controlling a technical system, relating pay to output becomes difficult. New remuneration systems have to be based on time worked, sometimes with additional payments for enhanced skills, and merit payments as a result of performance appraisal.

New information technology is expensive and often needs high levels of investment. To operate costly machinery profitably, there is an inherent pressure for it to be fully utilised. Often, this results in more shift work, weekend and overtime work. Thus, new information technology has an inherent potential to **increase the flexibility of time schedules**.

4. Information technology and participation: The strategies of the parties involved

In chapter I we have given an overview of the participation strategies of both unions and employers at the European level. The European Trade Union Confederation (ETUC) has taken a stand which might be called **participation as a democratic force**. Its main principle is involvement of unions and workforce representatives in all important decisions about the company, at the earliest stage and in co-determination procedures, if at all possible. The guiding principles were adequately expressed at the ETUC conference in Stockholm in 1988 (see chapter I). Such far-reaching aims are envisaged because of the pervasiveness of new technology on all aspects of the enterprise which can only be controlled by a far-reaching participation strategy.

This union approach to participation on the European level means a break with a long union tradition. So far, unions have been organisations that reacted to situations initiated by management. In this tradition, management prerogatives to decide important matters of the enterprise were not challenged. The unequal power balance between both parties was always seen and criticised. But this situation also guaranteed a wide union action field without too many restrictions to pursue the interests of the workforces. The new strategy of the ETUC means a distinct change, and it is an open question how far the separate national unions and workforce representatives are ready to comply with it. In many countries of the European community there is still a strong union tradition to refrain from involving trade union representatives in issues regarding work organisation, professional development and the like, let alone to be involved in matters of company policy. In many cases, they are afraid to become jointly responsible for organisational solutions which might bring short-term benefits to their workers but which might subsequently constrain their freedom of action and their ability to adequately represent their members' interests. In our survey, the Europe level analysis as well as the analysis of country differences will give evidence about attitudes of workforce representatives about their inclination towards these concepts of participation.

The participation strategy of the European Employers Association (UNICE) is less detailed than that of their union counterparts. As has been pointed out (chapter I) employers tolerate all participation procedures but stress the need that such practices should be "worked out by the parties directly concerned, according to local conditions, traditions and legislation". Thus, UNICE takes a stand against unitary solutions but stresses the need to take account of the circumstances of the particular country, industry or enterprise concerned. Given the historical and legal differences and differing industrial relations systems in Europe, they basically opt for an open situation of participation procedures. Yet, reviewing the management literature and social science research on workforce involvement in the process of recent technological modernisation, managers follow their own particular policies as well. Here, **participation is seen as a productive force**, and it is practiced basically as direct participation which fosters a culture of 'individualism', where individual employees or groups of employees are developed in line with the need of the enterprise.

Particularly in cases of flexible technology application and in search for optimal technical solutions, managers have recognised the essential contributions that workforces make to the successful implementation and operation of new information technology. As the European Foundation case studies have shown and which have been confirmed by many other studies, managers were aware that the workforce brings in considerable skills, experience and ingenuity in both adapting to and overcoming limitations in the technology. As a consequence, workers are often asked directly to contribute to technical change in quality circles, briefing groups, teamworking, or through other means. This trend for direct worker participation rises with the level of qualification in the enterprises, the complexity of technology applied and the uncertainty which this technology entails for the organisation. Managers often adopted policies to seek at least compliance but preferably the active commitment of the workforce in the process of technical change.

At present, such concepts are discussed under the heading of corporate identity and culture of the organisation. The guiding principle of these concepts is to extend a utilitarian engagement of workforces into a normative realm. Purely utilitarian engagement leads to problems of effectiveness as workforce control and motivation become precarious. Through creating corporate identity and a particular culture of the organisation, normative elements are added to the company, creating a 'we-feeling' and normative compliance of workers and employees towards the organisation. To achieve this, management is not only ready to directly involve workers but to grant them more autonomy as well, thereby cutting down direct control. The crisis of middle management testifies to the point.

The two strategies described are widely apart. The union strategy aims at representational participation on all levels while managers prefer direct involvement of the workforce if the situation necessitates it. But these strategies are not mutually exclusive. There has always been informal "participation" in enterprises at higher qualification levels. With the average increase of skill levels, the scope of such participation increases. There are also areas of trade union concern which cannot be covered solely by such procedures: employment and job loss, the core-periphery problem, the availability of training facilities for all workers, and many others. In addition, there is the problem of determining the course of technical change in which early involvement by workforce representatives might be helpful.

We have pointed out that technical solutions are often chosen by taking into account situational factors such as the shortage of qualified labour or the existence of powerful groups of employees with vested interests. Such technical solutions do not fully realise the growth potential of the technology, and as a result, diminish the profitability of the company, and opportunities for enhanced levels of pay for employees suffer. Early involvement of the unions in technical change might help to overcome such restrictions.

But such an approach to participation needs adaptation from both sides. Participation in the early stages of introducing new information technology cannot follow the traditional methods of identifying the objects of participation process primarily through a linear-type strategy - that is the forecast of cause-effect relations. Under conditions of uncertainty which these technologies create in enterprises, too many variables cannot be predicted, and a mechanistic cause-effect approach would only lead to a build up of oppositional position in an assent/dissent alternative which may divert attention from the critical issues. Thus, participation in early stages of technology introduction should basically involve an element of method, a highly flexible strategy that might allow the evaluation of both opportunities and risks as they are taking shape[6]. Such approach should be more concerned about participation as a procedure than as defining contents beforehand.

A possible approach to reconcile both concepts of participation as democratic and a productive force might be a model which is known as the positive-sum game. This is a social situation in which participants can attain their individual and sometimes differing objectives only through a certain degree of cooperation. If they cooperate their common situation improves, and there are net gains for all participants. If they fail to cooperate there is a possibility that both sides will lose. Such a positive-sum situation differs fundamentally from a zero-sum concept where the gain of one participant is always accompanied by an equivalent loss of the other side. Cooperation in such a situation would be detrimental to one's own interest. The strict pursuit of one's own interests at the other's expense is the only reasonable strategy in zero-sum games.

There are structural needs and constraints for both concepts. A zero-sum strategy is often found in situations when one party does not need the input and the cooperation of the other side over a longer period of time. If, for instance, employers only need unqualified labour and if there is abundance of low qualified personnel which can easily be substituted, the situation is not likely to result in cooperative relationships between employers and unions. Employers will seek to maximise their gains at the expense of the other side. In turn, unions will follow the counter-strategy of fighting for maximum gains for the workforce.

This was the typical situation in the early phase of industrialisation. Industrial relations in the manufacturing sector were largely antagonistic. Each side worked for short-term maximisation of benefits at the other's expense.

The situation in business life is completely different if both sides need each other and if they need each other over a longer period of time.[7] Under such conditions, both parties will gain through cooperation and participation, and both parties will lose it they attempt to pursue their own ends ruthlessly. In a situation of mutual dependence it is not advisable to work for short-term maximum gains, but to attain a long-term optimum. Such a strategy must not be mistaken as collaboration where one party gives up its own identity. Cooperation and participation do not mean the denial of differing interests. But any pursuit of one's own interests always takes into account the problems and constraints of the other side. Conflicts under such conditions are not intended to hurt the other side fundamentally, but to come to a fair compromise. Even if the gains of one side are higher in the short run, this might be accepted in the expectation that the benefits of such forms of cooperation will flow to both sides in the long run.

When we look at economies, enterprises and workplaces which operate with new information technology, the opportunities for such positive-sum behaviour of all parties involved are greater than any time before in West European economic history. New information technologies are technologies which develop rapidly, which have uncertain effects and which need continuing communication over their functioning and problems they create. In order to operate these technologies successfully and to achieve optimal results, organisations and the people who work in them have to be flexible. Cooperation and due respect to the interests of all those involved help to fully maximise the benefits which stem from these technologies. In such a situation, participation can combine the two elements of a productive and a democratic force.

NOTES

1. The introduction of robotics in Fiat's car division, for example, has radically upset the ratio between the traditional figures of direct and indirect workers. Over 70 percent of the staff is now made up of maintenance and regulation personnel whereas previously that same percentage represented direct workers concerned with manufacturing duties.
   Cf. P. Depaoli et al.: The Role of the Parties Concerned in the Introduction of New Technology, in: European Foundation for the Improvement of Living and Working Conditions (ed.): Participation in Technological Change (Consolidated Report), Dublin 1987, p. 35.

2. For a decription of different technologies cf. Appendix 8 of this report.

3. CAD/CAM: Computer-aided design/computer aided manufacturing.

4. A. Sorge et. al.: Microelectronics and Manpower in Manufacturing: Applications of CNC in Great Britain and West Germany, London 1983;
B. Lutz and P. Veltz: Maschinenbauer versus Informatiker - Gesellschaftliche Einflüsse auf die fertigungstechnische Entwicklung in Deutschland und Frankreich, in: K. Düll and B. Lutz (eds.), op. cit., pp. 213 - 285;
H. Hirsch Kreinsen, op. cit.

5. P. Cressey and V. di Martino: Workers' Participation in Technological Change, Dublin 1987, p. 47.

6. European Foundation for the Improvement of Living and Working Conditions: Participation and Technological Change (Booklet), Dublin 1987, pp. 38-40.

7. R. Axelrodt: The Evolution of Cooperation, New York 1984.

Chapter Three

Survey Methodology

The survey was designed to measure opinion of managers and employee representatives who were personally involved in the introduction of new technology which had a significant impact in the workplace. The methodology was agreed upon by a group of experts including representatives of unions and employers at a European level.

In the analysis of technology introduction and its results equal weight was given to the information provided by and opinions of both managers and the employee representatives. This was achieved by ensuring that pairs of managers and employee representatives of the same company or establishment were interviewed. The survey sample was restricted to larger establishments (at least 50 employees) which had some kind of formal employee representation and which involved these representatives in the process of technology introduction.

1. Countries and Sectors

The study covers all twelve member states of the European Community. Within these countries, all industries and services introducing new technologies were eligible for the study. For practical and methodological reasons, we had to limit the coverage of the survey:

- Telecommunications was left out for practical reasons. This industry is interesting in terms of new technology, but as it is highly monopolised and state-controlled in most countries, a survey would have posed unsurmountable practical difficulties.

- Car manufacturing as a major user of robotics and modern production control systems had to be excluded as it is not found in all European countries. Because of the lack of comparability, this major industry had to be left out of the survey.

In the end, the study concentrates on mechanical engineering and electronics in general to represent the industrial sector and on banking and insurance as well as retailing to cover the service sector. With these selections the main areas of new technology application are represented: Mechanical engineering covers a large and extremely differentiated area of activities which is established in all countries.

For mechanical engineering, electronics forms the technical base, and this line of production is hardly separable from the practical application in industrial production. In addition, electronics is considered to be a growth industry.

In the service sector, banking and insurance were chosen for two reasons: they were the first areas in the service sector to be heavily influenced by main-frame computerisation and they have experienced a second wave of new technology introduction with a new generation of more flexible computer technology. At the same time, banking and insurance also cover the vast area of word processing applications. Retailing is the second area of the service sector that was included in the study. It potentially shares the word processing applications with banking and insurance, but it is also of particular interest in terms of new developments such as store control, sales analysis or purchase order decisions which have been made possible by the new technology. Furthermore, retailing is traditionally characterised by a low degree of unionisation. Research as to involvement of employee representatives in technology introduction might produce interesting clues for future participation schemes and politics in service areas with low levels of unionisation.

2. Survey Procedure

The point of entry into the company was the personnel manager or his/her functional equivalent. A short telephone interview was conducted with this person to establish the company's eligibility in terms of sector, size, and existence of some form of employee representation. If the company satisfied our criteria for inclusion in the survey, the personnel manager was requested to name managers and employee representatives who were personally involved in the introduction process.

The potential respondents were subsequently contacted by letter or by telephone to arrange personal interviews. Checks were made to ensure that employee representatives had played a role in the introduction process.

To reflect the national situation, there are minor variations in each country as to survey organisation. Full details about the organisation of the survey are given in the Technical Appendix.

The survey procedure can be summarised as follows:

- telephone interview with the personnel manager to establish the eligibility of the company for inclusion in the survey and to gain cooperation,
- introductory letter sent to the potential respondents,
- personal interviews conducted with pairs of managers and employee representatives, preferably with two managers and two representatives of each company.

Through this procedure, a total of 7,326 interviews - with 3,663 managers and the same number of employee representatives - were carried out in the 12 countries of the European community. The interviews were carried out in two waves: between 11 February and 22 May 1987 in Denmark, Germany, United Kingdom, France and Italy; in Belgium, Greece, Ireland, Luxemburg, Netherlands, Portugal and Spain between 5 April and 7

October 1988. Altogether, interviewing took place in 2,807 European companies which means that - as a statistical average - 1.3 managers and 1.3 employee representatives were interviewed in each company. This is the largest survey to be conducted on the topic of employee representatives' involvement in technology introduction in Europe.

3. Limitations of the study: industrial climate and sectoral distortions

On the whole, there was a very positive response to the survey by both managers and employee representatives. However, two limitations of the study must be pointed out:

First, the findings of this survey must be evaluated in the context in which the interviewing took place, not only in terms of the survey design but also with respect to the social climate at the time of interviewing. In Germany and Italy, salary negotiations were conducted at the time of fieldwork (February - May 1987) and in both countries there was the prospect of industrial action in the mechanical engineering sector at the commencement of fieldwork. Not only were companies in this sector particularly cautious about allowing their staff to be interviewed[1], but the delicate state of industrial relations in both countries meant that companies in other sectors were also adopting a cautious approach.

Secondly, there were some difficulties as to sectoral representation in France and in the United Kingdom. In France, there was a high level of non-response in the mechanical engineering branch. Obviously, the introduction of new technology and the involvement of workers' representative in this process seems to be a particularly sensitive issue. The higher than average non-response rate can only be explained in terms of the historically tense industrial relations in the metal processing sector of this country.

Problems were encountered in the service sector as permission to interview at branch level had to be obtained from the headquarters. This proved to be a particular problem in the United Kingdom where two of the four main clearing banks refused to participate. The banking sector in the United Kingdom then experienced industrial action some months after the fieldwork period suggesting that the senior management at the time of fieldwork were anticipating some difficulties in the near future.

As to the countries in the second wave of interviewing, Ireland poses a special problem: very few establishments met the size or other criteria set for the survey. In spite of lowering of the standards of eligibility of companies, the number of interviews was particularly low in this country, the reason simply being the lack of establishments which were appropriate for the survey.

4. Limitations of the study: Screening procedure

The interpretation of the data has to consider some distortions which are due to the screening procedure of the companies and the respondents within the companies. According to these procedures, representative results in a strict sense are probably not achieved. The possible bias in the data can be worked out in three steps:

- The social partners, the Commission and the state representatives in the Working Group of the European Foundation agreed on a screening procedure that limited the eligibility of companies for the survey to firms which had formalised workforce representation. Because of this, companies without formalised workforce representation are systematically excluded. This first step in screening means that there is the possibility of over-representing in the sample companies which had a more positive climate of industrial relations.

- The point of entrance to the firm was the personnel manager. He or she decided whether or not interviewing could take place. In the knowledge that workforce representatives would also be interviewed, he or she might have made the decision to take part in the study as a means of assessing the general attitude of his or her counterpart as to the issue of participation. We might, therefore, expect firms with a positive climate of industrial relations to be over-represented in the sample, while conversely, companies with conflictual industrial relations might be under-represented. For the same reasons, it seems likely that companies with rather non-conflictual technology introduction, leading to positive results, are over-represented.

- According to the screening rules agreed upon, the personnel manager nominated a workforce representative to be interviewed. This procedure might have increased the chance for a biased selection of workforce representatives: personnel managers might have nominated counterparts who were closer to their point of view or who were not union members.

This possibility was tested by using data on union membership of workforce respondents in the survey and the results seem to indicate that - with the exception of one country - this assumption is not valid. In Germany, for instance, 60 percent of the respondents were members of the DGB-union, 12 percent were DAG-members and 23% were non-union members. In Italy we found only 8 percent non-union-respondents and the sample reflected the national proportions of membership in the three major Italian trade union confederations. Likewise, the Danish and British results do not suggest biased selection of workforce representatives as to union membership and non-membership.

Only in France do we observe major distortions: two thirds of the interviewees were non-unionists, and the relative shares of union representation are biased. According to a second indicator - the position of the respondent in the respective system of workforce representation - seems to confirm the deviant French result: 20 percent of the workforce representatives are classified as 'representant du personnel designé par le patron'; another 10 percent cannot be identified as to position. The German results are not biased in this sense: in more than 60 percent of the companies, the chairmen of the workers' councils were interviewed, and in another 15 percent of the companies the respondents were deputy chairmen of the councils. Only in 5 percent of the companies were non-council members interviewed. The Italian, Danish and British data as to the position of workforce respondents was seen to be unbiased, too.

Among the countries in the second wave of interviewing we face some distortions in two countries: in both the Netherlands and Spain, about four fifths of the employee representative respondents were non-unionists or refused to answer the question as to their union affiliation. Portugal marks the other extreme as virtually all respondents (97 percent) were unionised. With the exception of Belgium (70 percent unionised), workforce representatives in the remaining countries Ireland, Luxembourg and Greece indicate a medium degree of unionisation that ranges between 47 and 58 percent. All in all, representatives in countries which were interviewed in the second phase have less ties to unions than was the case of those interviewed in the first wave of interviewing. This might indicate a medium to rather strong bias in the selection of workforce respondents as to union affiliation in the second phase.

NOTES

1. For full details of the response rates cf. the Technical Appendix.

Chapter Four

Past Participation at the European Level

1. Introduction

As we outlined in the first chapter, the European Community has made workers's participation in the process of technological change a priority issue of industrial relations policy. The European Commission has adopted a role as moderator and facilitator for bridging the considerable gaps in attitudes and strategies that exist between the two parties directly involved at the European level, the employers' and the unions' organisation. The European Trade Union Confederation (ETUC), the union organisation, favours a participation model which implies very early union and workforce representative involvement at an early stage in company decision- making, possibly on a bargaining, co-determination type. UNICE, the employers' organisation accepts, in principle, all procedures of participation, but wants them applied according to country specific conditions, to local and to company specific situations. Although both strategies are difficult to bridge, both parties have reached a joint understanding on some fundamental principles: that technical modernisation of enterprises and workshops, workforces and their representatives should be informed and consulted in time, on issues relevant for the workforces affected.

For the first time, the European Foundation survey permits an inventory of actual participation practices in all EC countries. The survey results can help to clarify the situation of workforce representative participation vis à vis the strategies and intentions of both parties. In this chapter we want to answer the following questions:

- What was the actual extent of workforce representative participation in the European Community in 1987 - 1988? In particular, what is the extent of involvement in the different phases of technology introduction? Has there been participation in the early, strategic phase of technology planning?

- Apart from the phasing of involvement: has there been participation in substantive issues of company life? Again, this is a union demand. What was the intensity of such involvement?

- How do both parties perceive the extent, the intensity of participation in different stages and as to different issues? Are they in accord or do they disagree about actual levels of participation?

Data analysis to answer these questions will give the first clues as to how far the ideas and intentions of both parties are practiced already. For instance, are there indications that early participation in important matters like technology planning or other strategic company matters are an exception in Europe so far? Or are participation procedures mainly restricted to the later stage of technology implementation and to issues which are of rather secondary importance to the workforces and unions? Then, data about accord or disagreement of the two parties in regard to specific issues give important clues as to the climate in which both managers and workforce representatives are operating in. High dissent between managers and representatives as to the extent of participation in their enterprises and workshops would signal "low trust", antagonistic relations between both sides while high levels of accord between the two parties can be considered as an indicator for a rather co-operative style of management-employee representative relations in the companies.

At Val Duchesse, both parties came to a joint understanding that the information given by management should be timely to enable the other side to work out informed reactions. We will extend this particular aspect of the information giving to the broader scope of information quality. Timely information might be useless or too limited. Thus the following questions should be answered by the survey analysis:

- How do employee representatives evaluate information provisions? Does the information given conform to the standards of usefulness, completeness and early timing?

- In the process of technology introduction, does information giving follow in established, routine ways or have the parties involved created new channels of information?

The survey data permit these questions to be answered, thereby giving the first hints of how far the principles of the Val Duchesse Joint Opinion are already in effect and where we find considerable deviations.

In applying a European perspective in the survey we have to be aware that we are dealing with **widely differing notions of involvement**. Because of distinct historical, economic and political contexts, the term "involvement" has different connotations among the single European Community countries.[1]

Some important differences should be outlined here to give an impression about the complexity of the participation issue and to stress the need for empirical constructs which cover all possible notions of the concept and thus be able to serve as a common indicator.

- In the case of institutionalised forms of participation as they are analysed in this survey, the differences might be vast. Some countries rely more on contractual solutions between employers' associations and unions, whereas other industrial relations systems are based on law. Contractual forms of participation might be very centralised and be applicable nationwide or for whole industrial sectors while in other countries contractual agreements are made at company level only.

- The content of participation might vary considerably. Some countries rely more on agreements about procedures, others perfer to regulate concrete topics. Procedural approaches tend to regulate the form and process of workforce participation while substantive agreements centre around specific problems such as work organisation, safety regulations, qualification issues or remuneration.

When we talk about workforce involvement on the European level, we have to keep these diverse forms in mind. For an empirical approach, we need to grasp the notion of involvement in a way that allows for all these national differences to be taken into account. Therefore, the study adopts a very wide definition of involvement:

"Involvement in the introduction of new technology is defined as any participatory procedure or practice, ranging from disclosure of information to consultation, negotiation and joint decision-making, which formally or informally, directly or indirectly involves the parties concerned with the introduction of new technologies in the discussion of decisions concerning the process of change."[2]

According to this definition, involvement or participation within companies covers a wide range of possible activities, and any research faces the problem of formulating such terms more concretely. For this study, a hierarchical order of different forms of involvement has been adopted, its leading principle being their degree of intensity. By degree of intensity we mean the increasing opportunities of employee representatives to be drawn into managerial decision-making processes as to technological innovation:

No involvement: Management decides all alone. It plans and implements technology without any involvement of employee representatives.

Information: Employee representatives are informed by management about technological innovation in written form or in meetings. The scope of information and the mechanisms of information provision may vary greatly. They might include briefing sessions, information agreements or even group forums on change.

Consultation: There are joint committees, project or steerings groups in which employee representatives are not only informed but where they can voice their assessment of management decisions on technology introduction. In case of differing views, employee representatives can expect management to explain its decisions.

Negotiation and joint decision making[3] are the most intense forms of workforce participation: binding results as to technology introduction are worked out between the parties. These bargaining mechanisms can be based on nationwide legal provisions, on productivity or new technology agreements between the social partners on a nationwide, sectoral or company level.

Thus, the degree of intensity of involvement can vary between the extremes of "no involvement" and "negotiation/joint decision making" in an ascending order. It is evident that the ability of employee representatives to influence the form and direction of new technology increases with the rising level of involvement, with information provision being the minimum of positive involvement procedures.

2. Extent and timing of past involvement

The introduction of new technologies must be considered a process that stretches over a considerable period of time. Such a process can be viewed as consisting of the following phases:

Planning phase: At this early point, fundamental decisions on technological innovations are made. They specify basic company goals as to what might be achieved by introducing a certain technology and how it should be achieved.

Selection phase: On the basis of decisions made in the planning phase, the appropriate technology is selected.

Implementation phase: The selected technology is installed in the company. This phase is problematic insofar as the new technology has to be integrated in the existing organisation.[4]

Very often new forms of organisation have to be developed. Because of the novelty and the potentially systemic consequences of the new technology with repercussions not only for the particular department, but for other departments and business functions as well, existing forms of organisation often turn out to be inadequate. Therefore, the implementation phase can be characterised as an experimental stage. At the same time, production has to continue and services have to be rendered. For a certain period of time, both the old and the new technology exist side by side until the new methods and the new organisation are foolproof.

Post-evaluation phase: In this phase, the technology is installed. Work organisation is adapted to accommodate the new technology, teething problems are eliminated, and the technology undergoes review in the light of the original goals.

Viewing the four phases in perspective, the planning and the selection phase may be characterised as the **strategic**, the implementation and the post-evaluation phase as the **operational** phases of technological innovation. The first two phases are strategic insofar as fundamental decisions are made here which to a large extent predetermine production structures and work organisation. Activities and decisions in the operational phases take place within a rather restricted frame of reference, where fundamental decisions have already been made in the strategic phases. Decisions and actions in the operational phases are important for the company, but in comparison to the strategic phases, they are of lesser significance.

In each of the four phases workforce involvement is possible. We would normally expect a greater degree of participation in the operational phases: The strategic phases are more important for managerial functions and prerogatives, and it is reasonable to expect that management is hesitant to have its prerogative challenged by workforce involvement. In the operational phases, participation has consequences, too, but they are of a minor nature as the more important decisions have been taken earlier. In addition, management might seek workforce involvement in the operational phases to guarantee the smooth functioning of the new technology and any consequent work organisation.

In the interests of clarity and the need to economise in the display of data, we concentrate on the most important strategic and operational phase, on the stage of technology planning and introduction. This decision heightens as well as lowers the overall rates of workforce involvement as they are included in the data: the results suggest that workforce involvement is markedly lower in the selection phase than in the planning phase. Obviously, technology selection needs highly qualified expert knowledge and it is the domain of technical experts. In general, workforce representatives' expertise is too limited to be effectively involved in such processes. Leaving out participation rates in the selection phase mean an average increase in the involvement rate in the strategic stage which "involvement in the planning of technology" stands for. In the operational phase, participation rates in the post-evaluation stage of technology introduction are highest. By concentrating on the preceding implementation process we are dealing with a lower rate of workforce involvement which gives the operational phase a more negative character. Yet, considering the limited aspects which are touched upon by workforce involvement in the final stages of implementation and which can be characterised just as technology assessment on the company level, participation at this stage of the whole process should not be regarded as a very important political asset.

Figures 1 and 2 hold the data on the intensity of workforce participation in both the planning and the implementation phase of technology introduction, separate for management's and workforce representatives' view.

At the European level, workforce participation in the phase of technology planning is rather limited. According to managers' assessments (fig. 1) there is no workforce involvement in two fifths of the cases. An equal share of the management respondents report the provision of information as the only means of drawing workers and employees into the strategic process of technology planning. The remaining share is almost equally distributed between consultation procedures and negotiation/joint decision making (10 percent each).

In the phase of technology implementation, involvement of workforce representatives is much more prominent. Situations of no involvement have decreased by almost one half (to 22 per cent). Information giving holds an equal share of 39 per cent which might indicate a stable pattern of workforce involvement on a low level of intensity in two fifths of all cases. But the more elaborate forms of consultation and negotiation and joint decision procedures have almost doubled in comparison to the planning phase: about one fifth of the managers report consultation practices in their companies and an almost equal share have negotiated solutions and have jointly decided on problems arising in the process of technology implementation.

On the whole, employee representatives assess the situation in similar terms as their management counterparts. As a trend, they are slightly more critical as to the occurrence of the higher intensities of participation, and they report a higher rate of no involvement in both the planning and the implementation phase. But the differences between both social parties are minimal in the end, and the more **significant result is the broad agreement in the answers of managers and workforce representatives**. From a methodological point of view, such a high consensus among sub-groups is very unusual in attitudinal surveys. We therefore have to judge these results not just as opinions regardless of facts but as a valid description of a genuine situation.

In reflecting upon these results in the framework of the Val Duchesse Joint Opinion of both the European union and entrepreneurial organisations, the reality of participation in European firms applying information technology is in marked contrast to the intentions of both parties. If the message of Val Duchesse is that the workforces and their representatives should be informed and consulted about various aspects of technical modernisation, the survey data shows that this aim is not achieved in the majority of European enterprises. This particularly pertains to technology planning where only every fifth respondent indicated participation practices of a consultative or a co-determination type. Moreover, in two out of five companies there was no indirect workforce involvement whatsoever. The reality of participation looks somewhat more positive in the less important, operational phase of technology implementation where two fifths of the managers and one third of

Figure 1: Past participation of employees in planning and implementing new technology in the EC countries - Managers

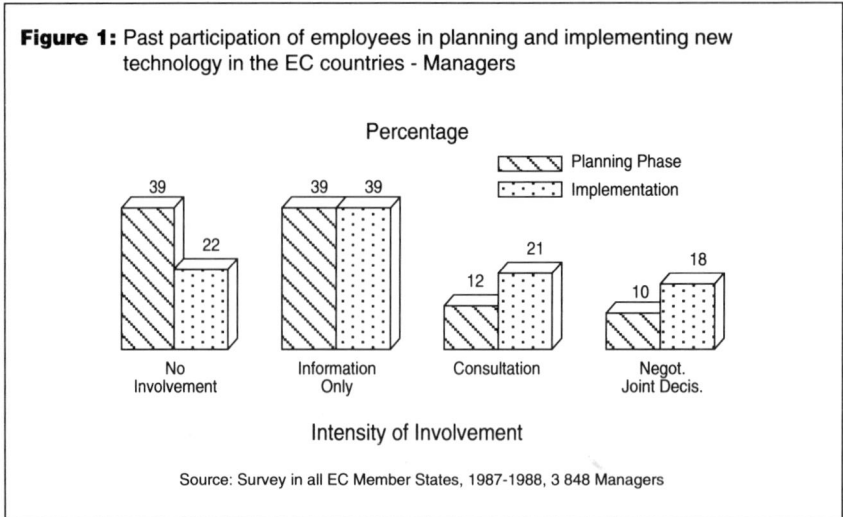

Percentage

Planning Phase
Implementation

Intensity of Involvement	Planning Phase	Implementation
No Involvement	39	22
Information Only	39	39
Consultation	12	21
Negot. Joint Decis.	10	18

Source: Survey in all EC Member States, 1987-1988, 3 848 Managers

Figure 2: Past participation of employees in planning and implementing new technology in the EC countries - Employee Representatives

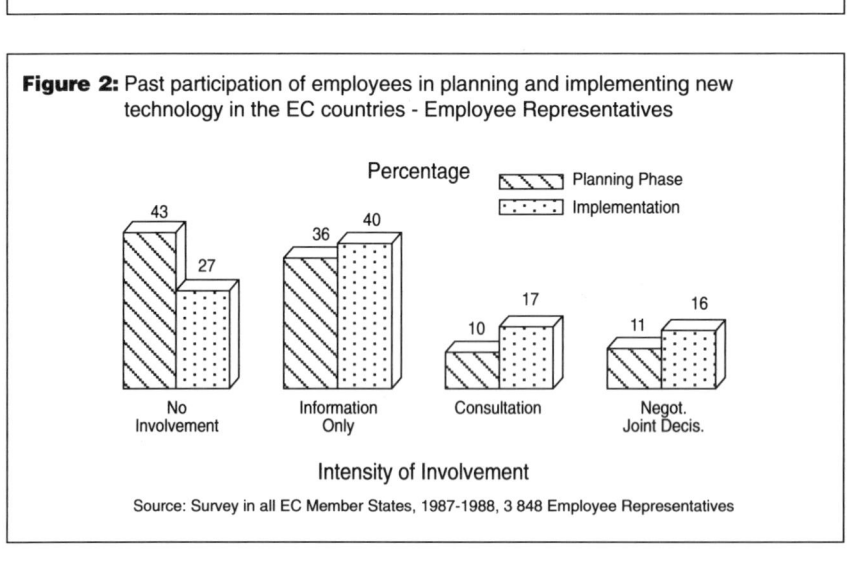

Percentage

Planning Phase
Implementation

Intensity of Involvement	Planning Phase	Implementation
No Involvement	43	27
Information Only	36	40
Consultation	10	17
Negot. Joint Decis.	11	16

Source: Survey in all EC Member States, 1987-1988, 3 848 Employee Representatives

employee representatives report consultation and co-determination practices in the past. This is at the expense of no involvement. In this operational phase, simply informing the workforces about technical changes remains the most frequent single means of involvement in the planning phase of technology introduction - a policy which is clearly not in-line with what both sides had agreed at the European level. Thus, the reality of participation in Europe falls way behind the intentions of both sides.

But a positive result should also be highlighted: both managers and workforce representatives are in high accord about the intensity of participation practiced in their enterprises so far. At a factual level both sides are in accord. These equal perceptions can be interpreted in two ways: first, this result indicates that facts have been measured in the survey, not just opinions unrelated to reality. Such high agreement is rare in social science and underpins the quality of the data. Further, and equally important, these similar perceptions can be taken as an indication that the relationship between both sides in the enterprises is mainly positive and co-operative. In case of antagonistic relations, we could expect rather differing views even on facts. But this is not the case in the European companies under study, and the presumed co-operative relationship between both sides points to a general "high-trust" atmosphere between both sides, a precondition for participation procedures which was asked for in the Val Duchesse Joint Opinion.

3. The content of past participation

The findings so far reveal that participation in the strategic planning phase can be characterised by a combination of no workforce representative involvement or mainly information provision as the lowest intensity of involvement. In the operational phase of technology implementation the pattern of involvement is more balanced and includes sizeable shares of consultation, negotiation and joint decision making. In other words: participation is rather restricted in a domain which by tradition is a management prerogative, namely the planning of organisational functions. The closer we get to workplace problems and to specific topics in regard to technology implementation, the higher the forms of workforce involvement.

This pattern which refers to the timing aspect of participation recurs when we investigate the content matters of workforce involvement.

In figures 3 and 4, eight substantive issues are grouped into management and workforce concerns. Such a wording needs commenting as it might create a false connotation. Of course, all these issues are of concern to both sides. But there are differences nevertheless as this grouping does

Figure 3: The content of past participation - Managers

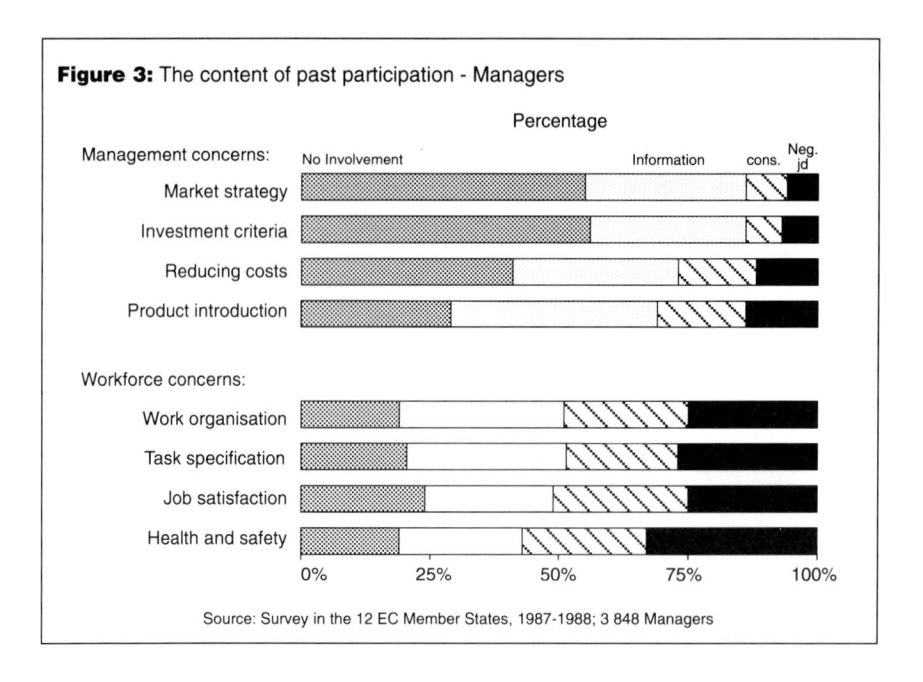

Percentage

Management concerns: No Involvement — Information — cons. — Neg. jd
- Market strategy
- Investment criteria
- Reducing costs
- Product introduction

Workforce concerns:
- Work organisation
- Task specification
- Job satisfaction
- Health and safety

0%   25%   50%   75%   100%

Source: Survey in the 12 EC Member States, 1987-1988; 3 848 Managers

Figure 4: The content of past participation - Employee representatives

Percentage

Management concerns: No Involvement — Information — cons. — Neg. jd
- Market strategy
- Investment criteria
- Reducing costs
- Product introduction

Workforce concerns:
- Work organisation
- Task specification
- Job satisfaction
- Health and safety

0%   25%   50%   75%   100%

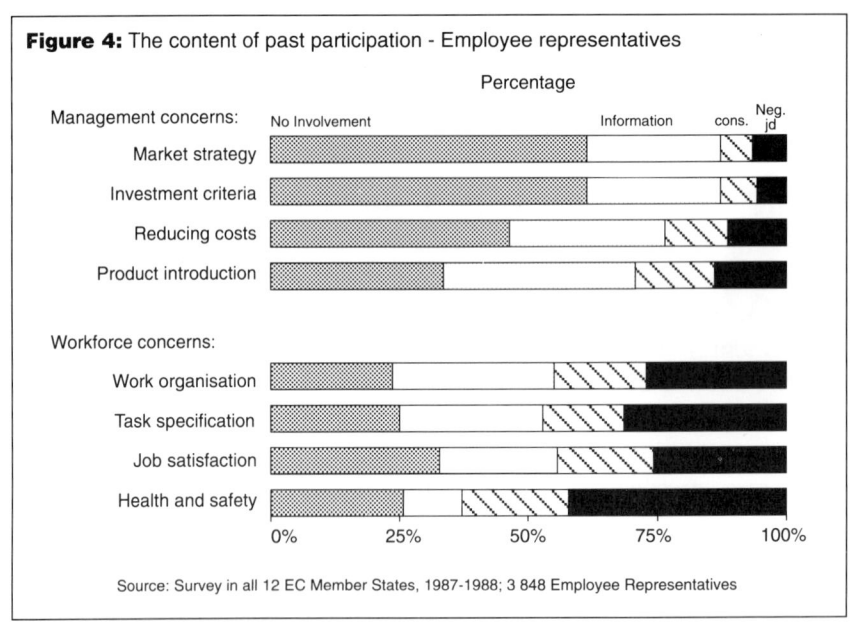

Source: Survey in all 12 EC Member States, 1987-1988; 3 848 Employee Representatives

express the traditional priorities of managers and the workforces. From a management point of view concerns regarding markets, stategy, investment, new products and costs are of primary interest as they form the core of management functions. For employee representatives, these issues are also important, as decisions taken here have repercussions on the functioning of the whole organisation. But given the defensive role of traditional workforce representation, these issues are of secondary importance for the representatives. Their primary concern is to minimise the negative repercussions on workers and to find positive solutions to developments which stem from strategic management decisions. Thus, what are primary management concerns are secondary issues for workforce representatives and vice versa.

The data in figures 3 and 4 reveals a pattern of participation which, in principle, reflects the idea of primary and secondary interests of both parties. It indicates that participation in substantive matters follows, as a trend, the traditional demarcation lines. In the area of genuine management functions, particularly in the basic issues of investment and market strategy, workforce participation is very rare. More than every second manager indicates that there is no workforce involvement at all in his company. Employee representatives report an even worse situation:almost two out of three indicate that as to these issues there has been no participation in their companies. If there is involvement it is mainly restricted to being informed by management. Consultation and co-determination practices in the area of primary management concerns are extremely rare. According to both managers and employee representatives these practices have been applied in about 12 - 13 % of all companies in regard to market and investment strategies, with almost equal shares of consultation and negotiation/joint decisions. In regard to the less fundamental issues of cutting costs and introducing new products, both sides report a doubling of these shares; but they remain low nevertheless. Thus, where core management functions are at stake, involvement of workforces is very low. This situation can be characterised first of all by no involvement, followed by information procedures.

The pattern of involvement looks distinctly different in the case of primary workforce concerns. In issues of traditional workforce interests like work organisation, task specification, job satisfaction, health and safety, not to involve employee representatives has become a rather rare practice which is found in about 20% of the cases of the companies. Although there are still sizeable shares of information only, consultation and well as negotiation and joint decision making procedures are very frequent. As to work organisation, task specification and job satisfaction, every second manager indicates indirect workforce involvement in the way of consultation and codetermination. Employee representatives report somewhat lower shares, but in regard to health and safety measure, both parties are in high accord that in more than 60% of the cases consultation and bargaining procedures have been applied in the past.

The data about indirect employee participation in workforce concerns are certainly positive when compared to the equivalent data in regard to management concerns. But at the same time we must not overlook the fact that even in matters of working conditions - issues of immediate concern to workers and employees - in every third European company workforce representatives are only informed about changes due to technology introduction, and in every fifth European company, even this information is not provided. These results throw a shadow over the internal conditions of quite a few European enterprises.

The data about participation in substantive matters of company life resembles rather closely the result about involvement in the two phases of technology introduction. In the strategic planning phase and in management concerns, the situation is best characterised as a mixture of no and low participation of workforce representatives. In case of the operational phase of technology introduction, and in regard to secondary management interests, more participation opportunities are evident, up to co-determination practices. Thus, **management prerogatives are rather safeguarded in the process of technological modernisation, while the involvement of workforce representatives is still mainly restricted to secondary matters**.

4. The information issue

Participation is feasible only if all parties involved have sufficient information on the topics which have to be discussed and solved. Sufficient information is the base of all involvement. Without it participation processes are necessarily ineffective, and the whole notion of participation is futile. It is this basic understanding which induced the social parties in the Val Duchesse Joint Opinion to be very specific on the information issue related to processes of technical change in companies. In this joint understanding, they define information as "the action of providing workers and/or their representatives, at the level concerned, with relevant details of such changes, so as to enlighten them as to the firm's choices and the implications for the workforce." They further stress the "need to provide timely information concerning major changes in the terms of employment and working conditions of the staff concerned...".

These provisions point to four aspects of the information issue:

1. Technical change is initiated by management. Its choices have repercussions for the workforce which means that the workforces and/or their representatives are in a passive, reactive role. Thus, the information issue is salient first of all for workers, employees and their representatives. In the process of participation, they are dependent on information from management. Accordingly, in the scope of this analysis, the information issue is to be investigated from the **point of view of workforce representatives**.

2. Both parties have agreed on **timely information** provision indicating the fact that participation procedures are most effective in situations where decisions still leave some room for workforce intervention. Given management's prerogatives to make the basic decisions on technical change, early information provision helps to mould the repercussions of these decisions. In case of very early information provision, even the decisions themselves can be influenced which might lead to very different consequences for employees.

3. The Val Duchesse Joint Opinion holds implicit restrictions when **information giving should be restricted to the workforce "level concerned"**. In principle, this limits the scope of recipients of information on the workforce side which might create irritation and be a source of conflict. The parties might disagree about the proper potential recipients and about the scope, timing and usefulness of the information given for these recipients.

4. Taking up the concept of **usefulness**, information provision might be applied in a way which serves the concept of involvement only nominally while, in fact, useless or irrelevant information is supplied. In such a case, this would mean an erosion of the participation issue.

Data analysis so far has shown information provision to be the most frequent means of participation on the European level: this is so in both the planning and operational phase of technology introduction as well as in all primary management concerns. It is also true regarding two issues of primary workforce concern,[5] work organisation and task specification. These findings were treated as the facts of past involvement. They have to be supplemented by the additional dimension of how workforce representatives evaluate the facts of information provision in the framework of topics and problems discussed above, namely the scope, timing and usefulness of information given.

Figure 5 conveys an **equivocal picture of the information issue**: Workforce representatives evaluate the practice of information provision in positive as well as in negative terms, with a **slight trend for a positive assessment**. The usefulness of information is very positively assessed

Figure 5: Quality of information

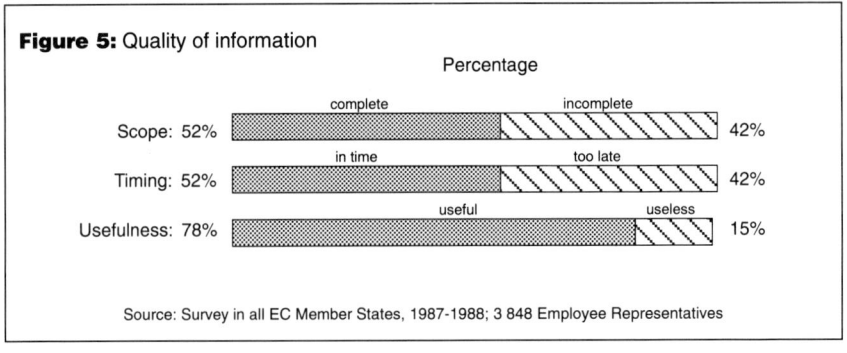

	Percentage	
Scope: 52%	complete / incomplete	42%
Timing: 52%	in time / too late	42%
Usefulness: 78%	useful / useless	15%

Source: Survey in all EC Member States, 1987-1988; 3 848 Employee Representatives

Figure 6: Overall evaluation of the quality of information by employee representatives

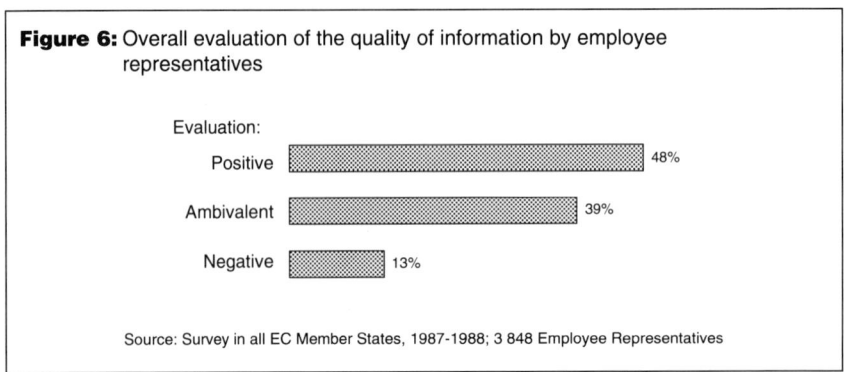

Evaluation:
- Positive: 48%
- Ambivalent: 39%
- Negative: 13%

Source: Survey in all EC Member States, 1987-1988; 3 848 Employee Representatives

by management: about four out of five respondents regard the information they received as useful. The other two aspects of the information issue are less well evaluated. Every second workforce representative has a similar positive attitude toward the scope and the timing of information. But as to these aspects, there are distinct criticisms as well. Two out of five respondents are dissatisfied with the amount of information they received so far, and a similar number feel that the information came too late. As "information" is the most important single participation measure on the European level the fact that many consider the information they receive to be incomplete and late casts a shadow over the situation of involvement in the process of technology introduction in Europe.

To carry the analysis further, we develop a unified measure for the quality of information. This measure should answer the question of possible overlap of positive or negative evaluations of the time issue: Is there an identical pattern of complete, rightly timed and useful information? Does incomplete, late and useless information cluster as well? In figure 6 all three evaluation dimensions have been combined into such a measure.

The wording of the four types of evaluation in figure 6 needs some explanation. A positive assessment of the information issue means that workforce representatives consider the information they received from management in the past to have been complete, timely and useful. **Half of all the workforce respondents in Europe describe their past information situation in such unequivocal positive terms**. The experience of two out of five employee representatives (39 percent) is one of ambivalence: the information they received was neither timely nor useful; nor was it complete enough to justify an outright positive assessment. **In 13 percent of the cases information provision was negative** on all three counts: it was neither useful nor timely nor complete. When we compare these survey results with the standards agreed in Val Duchesse, the situation at the European level elicits an ambivalent assessment. A share of 13 percent clear negative experiences of workforce representatives about the quality of information given by management is certainly not alarmingly high.

However, we must not forget that in another 40 percent of the cases representatives have been confronted with information practices which were generally inadequate. Satisfactory information was experienced by only half of the employee representatives in the enterprises under study.

The Joint Opinion reached in Val Duchesse points out the existence of various forms of information, consultation and negotiation practices in the member states of the European community, and workforce involvement is understood to take place within the framework of such nationally different provisions. The diffusion of new information technology in Europe has meant that both sides have to adapt to a rapidly changing environment, as was mentioned earlier in the theoretical part of this report. The higher degree of uncertainty for all participants in the process of introducing modern information technology means there is a need for the active support, motivation and technology acceptance by employees. Thus, existing and well-established channels of involvement have sometimes been supplemented by additional procedures and practices to smoothe the processes of industrial transformation and to make them effective and socially acceptable.

So far, we have no knowledge about the frequency of additional regulations which have been applied in Europe to help solve the information problem of participative processes. Figure 7 (page 58) gives a first overview over established and new channels of information provision on the European level:

Figure 7 reveals that in the vast majority (83 percent) of cases the information problem within companies is handled in a fashion which is in line with traditions of the respective countries. As these traditions differ considerably from country to country we cannot draw any positive conclusions here as to what these channels might be in detail and how they differ between the European nations. What we can conclude is that

at the European level the flow of information about new technology within companies mainly takes place in established channels.

New forms of technology specific solutions are rather rare: only 17 percent of workforce representatives report special arrangements; such arrangements are predominantly ad hoc and are not based on formal measures. According to figure 7 two thirds of the special arrangements regarding the flow of information concerning technological modernisation are ad hoc measures: the parties reacted in a flexible way on information needs when particular problems arose with the companies. Only 20 percent of these arrangements were negotiated in companies or between the social parties on a sectorial or nationwide level which resulted in contractual agreements. Special legal provisions to ensure an adequate flow of information were indicated by 13 percent.

Particular arrangements for the flow of information lead us expect a higher quality of information as compared to normal routines. As figure 8 shows, this is hardly the case.

Figure 7: Channels of information on technology introduction according to employee representatives

Normal channels 83% / Special arrangement 17%

ad hoc 67%
contract 20%
law 13%

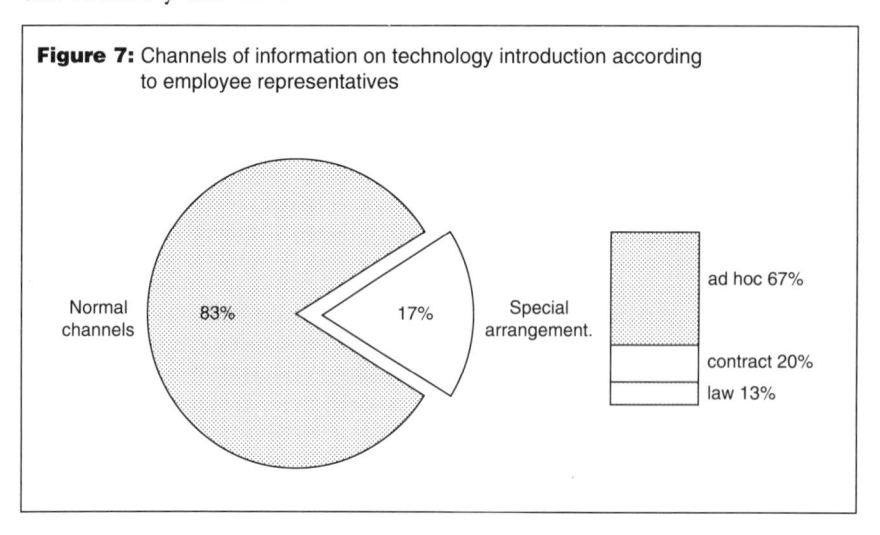

Figure 8: Evaluation of the quality of information dependent on the channels of information - Employee representatives

Evaluation:	Normal channels	Special Arrangements
Positive	48%	53%
Ambivalent	40%	40%
Negative	13%	7%

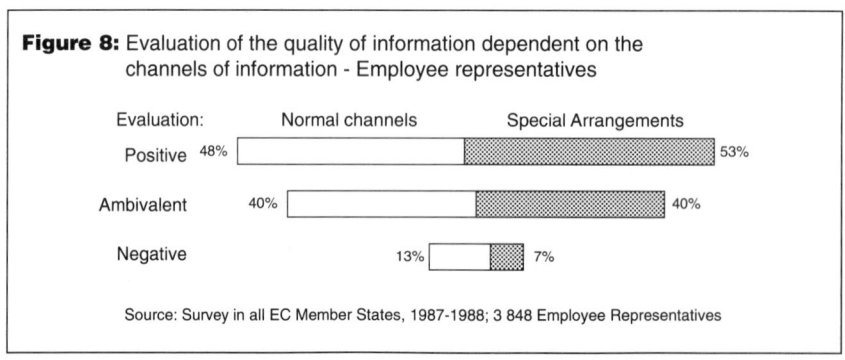

Source: Survey in all EC Member States, 1987-1988; 3 848 Employee Representatives

The pattern of **overall evaluation of the quality of information is basically the same whether such information provision is based on routinised procedures or on special arrangements**. There is little evidence to suggest that special arrangements have more positive results. This can be seen when looking at the two extremes of evaluation: The positive assessment under conditions of special arrangements are more frequent (6 percent) than under normal conditions, while both approaches generate exactly the same shares of ambivalent evaluations (40 percent). On the whole, both approaches function in a similar fashion.

In regard to the effectiveness of the different special information arrangements, there seems to be a certain trend.

Of all three approaches, **ad-hoc measures for information provision seem to be more effective than contractual and legal agreements**: the positive evaluations of information quality are highest (54 percent) in the case of ad-hoc solutions, followed by contractual solutions (49 percent). Under conditions of legal arrangements, only 38 percent of workforce representatives express an outspoken positive attitude. A restricted quality of information which was characterised as "ambivalent evaluation" is less often found in the case of ad-hoc measures and most often found under conditions of legal arrangement. Although the data

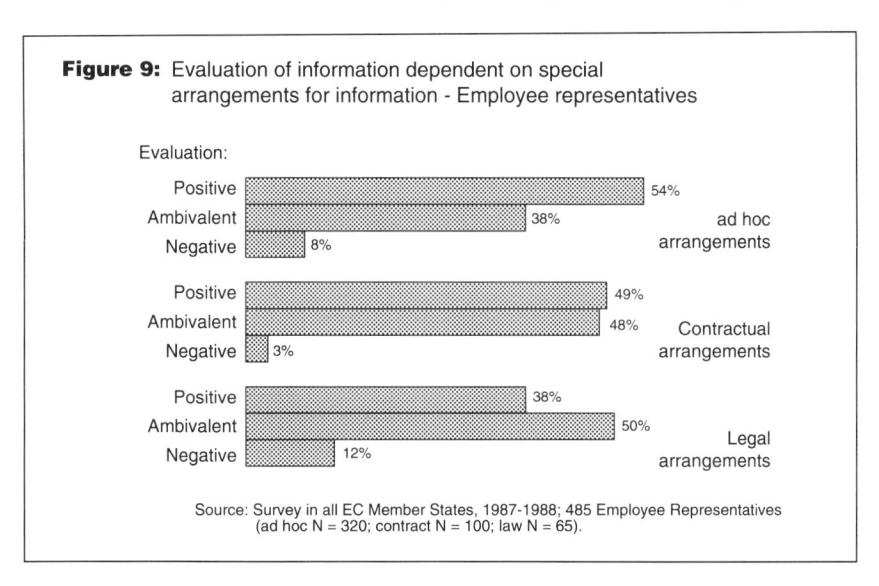

Figure 9: Evaluation of information dependent on special arrangements for information - Employee representatives

Evaluation:

ad hoc arrangements
- Positive: 54%
- Ambivalent: 38%
- Negative: 8%

Contractual arrangements
- Positive: 49%
- Ambivalent: 48%
- Negative: 3%

Legal arrangements
- Positive: 38%
- Ambivalent: 50%
- Negative: 12%

Source: Survey in all EC Member States, 1987-1988; 485 Employee Representatives (ad hoc N = 320; contract N = 100; law N = 65).

base for this analysis is not ideal[6], such a result is in line with theory and international experience which indicates that flexible solutions are particularly appropriate when dealing with new information technology (see chapter II). If this particular result is empirically sound it would convey the message that the more flexible provision of information has a greater chance of producing positively evaluated solutions than more rigid approaches. This would strenghten the idea to favour procedural-type participation solutions at the expense of substantive involvement procedures. But we do have to be aware that this result pertains to situations and companies where information provision is already a standard practice. It does not apply to the more serious problem of providing information at all.

In sum, half of workforce representatives assess the problem of information quality in the process of technology introduction in very positive terms. Very negative experiences are voiced by a minority of 13 percent only. This pattern of evaluation does not vary whether information is provided through established, traditional channels (which is the norm) or via special arrangements for information. But in distinguishing such special solutions to the information problem, flexible approaches have yielded somewhat more useful information than rigid solutions. In regard to the Val Duchesse joint opinion, two types of statements are possible: there are still **considerable gaps to be filled to reach the agreed standard of information provision** in European enterprises. Further, to achieve a level of high quality information about technical change, the parties involved should seek to **employ additional flexible means for information provision when new technology is being introduced into enterprises.**

5. Conclusions

When we review the situation of indirect, workforce representative participation in new technology in European companies in 1987-1988 in a political perspective, the following conclusions are appropriate:
At the time the survey was carried out, the social partners at the European level had agreed at Val Duchesse that information and consultation should be the norm of indirect workforce involvement in European companies. The survey data reveals a cleavage between these stated intentions and the reality in European enterprises. This cleavage is particularly obvious in strategic management interests, in the planning of new technology and in genuine management functions like defining market strategies and investment criteria. As to these issues, the situation in Europe can best be characterised as a combination of no indirect workforce involvement and information only. The common intention of both parties, to inform and consult employee representatives in the process of technology introduction and application is clearly out of line with reality when it comes to these strategic matters.

The facts of workforce participation are somewhat closer to the intentions of both sides when regarding issues in which management has less vested interests. As to implementing new technology, the Val Duchesse benchmark of at least consulting workforce representatives is achieved in every third European company under study, and in every second company there was consultation and co-determination in issues traditionally related to employee concerns. To express the result negatively: on the average, the stated intentions of both sides to inform and consult employee representatives were not achieved even in case of operational issues in the scope of technology introduction.

Thus, the survey analysis shows that management prerogatives were widely maintained. In the European community at the time of 1987-1988, there was a considerable gap between the reality of indirect workforce participation at company level and declarations of both sides at the level of European organisations. In spite of this considerable cleavage, the relationship between both management and the workforces does not seem to be disturbed. There was broad agreement with both sides within the firms. If there were antagonistic relations we could have expected dissent even as to the facts of past participation. But this is not the case. This result puts a first question mark behind the stated intentions of the European unions when they advocate a participation model of far reaching industrial democracy in the future: the reality of participation is very distant from such aspirations, and given the overall positive relations of both sides at company level, such a participation model is hardly likely to be achieved in the foreseeable future.

NOTES

1. Cf. chapter VIII for a detailed discussion of country specifics.

2. Manfred Grunt: The Involvement of the Parties Concerned in the Introduction of New Technology. Attitudinal Survey. Working Group Report, Munich 1986, p. 8.

3. In the five country study which preceded this report and which covered some issues analysed here, a different coding was used for data analysis: consultation and negotiation were grouped as one category while joint decision making was used separately to connotate co-determination practices (cf. D. Fröhlich, D. Fuchs and H. Krieger: New Infornation Technology and Participation in Europe: The Potential for Social Dialogue, Luxembourg 1989). This former coding is not used in the present analysis: both negotiation and joint decision making are treated as basically comparable procedures which lead to binding results for the actors involved. Accordingly, they are grouped in one category.

4. The situation is different in cases where new firms and/or in new sites are set up ("green field sites"). They do not pose problems of restructuring the organisation, but often have other problems to solve, such as single union/no strike agreements which is a source of great controversy within the British trade union movement.

5. As to job satisfaction, consultation procedures are slightly more prominent, and in regard to health and safety, co-determination is the participation measure most often exercised.

6. The number of respondents indicating contractual and legal arrangements is rather small.

CHAPTER FIVE

The Perceived Effects of Participation

The involvement of workforces and their representatives in the process of technological modernisation is not an end in itself; it is a policy instrument of which the parties concerned expect outcomes which support their own interests. They might try to avoid negative developments through participation or to achieve positive goals - both seen from their own point of view. It is natural that past experiences regarding the process of participation and its outcome shape the general climate and the opportunities for the participation generally. Negative effects of participation will place a question mark on the whole idea, while visible benefits will create an atmosphere for the continuation and extension of a policy which has proven its worth.

On the basis of the survey data available we want to investigate the topic of perceived effects of participation in four ways:

- to give an overall impression on selected effects of workforce representative participation,
- to find out in what way the perceptions of both parties differ or are in accord,
- to find out sensitive and problematic topics of participation,
- and to ask what intensity of participation - from information to co-determination - yields positive or negative results.

As a starting point - and as a base for analysing these topics - let us summarise briefly the experience both parties have had with the process of participation so far:

- As to the strategic planning phase, the more binding forms of involvement like consultation, negotiation and join decision- making were fairly rare.
- These demanding forms of participation were more prominent in technology implementation, in the operational phase.
- Similarly, participation in management concerns is rather restricted. Workforce representatives are largely excluded from management's considerations and decisions about basic company problems.
- As to workforce concerns, there is a surprisingly high level of co-determination in some of the companies under study.
- Information is the single most prominent means of involvement in both phases of technology introduction. The quality of this information is criticised by every second employee representative.

The overall impression which we gain of participation practices in high tech companies in the European community is one of ambivalence: In strategic company matters the situation can be characterised as a combination of no involvement at all or low intensity of involvement of workforce representatives whereas in operational problems there is a clear trend towards more intensive participation practices. Against this background we now analyse the issue of the effects of participation in three broad content dimensions:

- In which way does participation enhance or impair the process of technical modernisation of companies? This problem is analysed in regard to the practical issue of possible delay or acceleration in decision-making and in technology implementation.

- What is the impact of participation on the mutual understanding of the two parties in the firm? Does participation provide a base for better understanding between the two parties, or is it only a means for the institutionalised change within the intra-company power balance which might lead to increased conflict in the long run?

- Does participation improve the acceptance of new technology and does it ease technological innovation?

These topics are embedded in the wider frame of reference as discussed in chapter II: can the effects of participation be categorised under one of the three strategic approaches of participation, namely participation as a productive force, as a democratic force or as a positive-sum game?

1. An overview of the costs and benefits of participation

Figures 10 and 11 display the perceived effects of workforce involvement in nine dimensions which were presented to the respondents[1]. These dimensions are ordered according to their proportion of positive effects.

Three main results as to the effects of participation stand out:

- In the majority of effect dimensions, workforce involvement had neither positive nor negative effects in the companies.

- On average, negative effects are reported extremely rarely. The positive changes clearly outweigh the disadvantages of participation procedures.

- In general, managers see slightly more positive changes than employee representatives; but the differences between the two parties are very small.

A more detailed evaluation of the results reveals some interesting differences in the evaluation of certain effects which are discussed from a management point of view first. Three comprehensive problems are discussed: participation and the problem of time constraints, the development of industrial relations, and participation in the context of workforce identification with the company and technology acceptance.

Figure 10: Effects of past participation - Managers

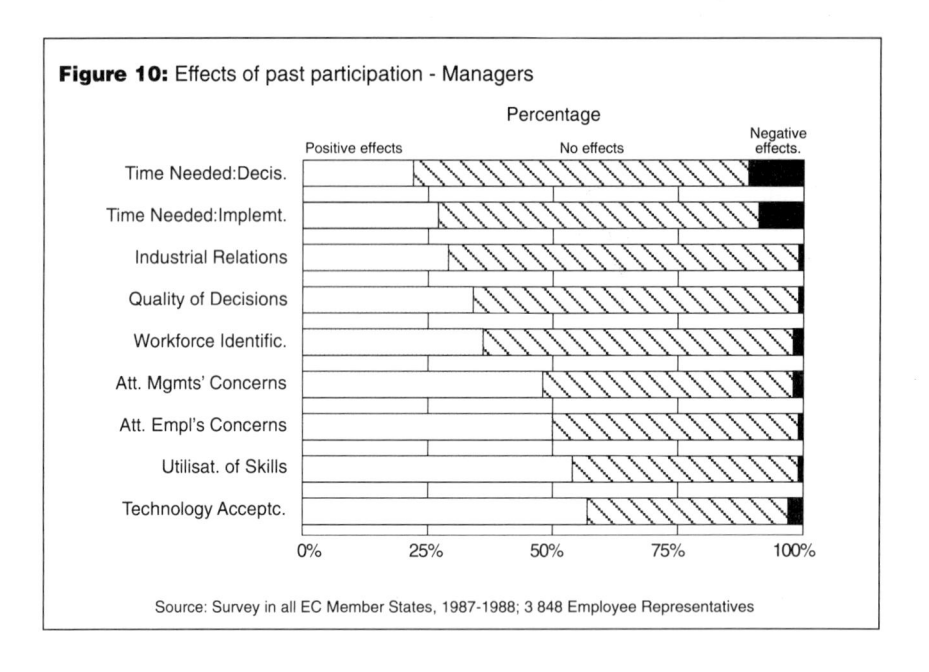

Categories (top to bottom): Time Needed:Decis., Time Needed:Implemt., Industrial Relations, Quality of Decisions, Workforce Identific., Att. Mgmts' Concerns, Att. Empl's Concerns, Utilisat. of Skills, Technology Acceptc.

Legend: Positive effects | No effects | Negative effects.

Source: Survey in all EC Member States, 1987-1988; 3 848 Employee Representatives

Figure 11: Effects of past participation - Employee Representatives

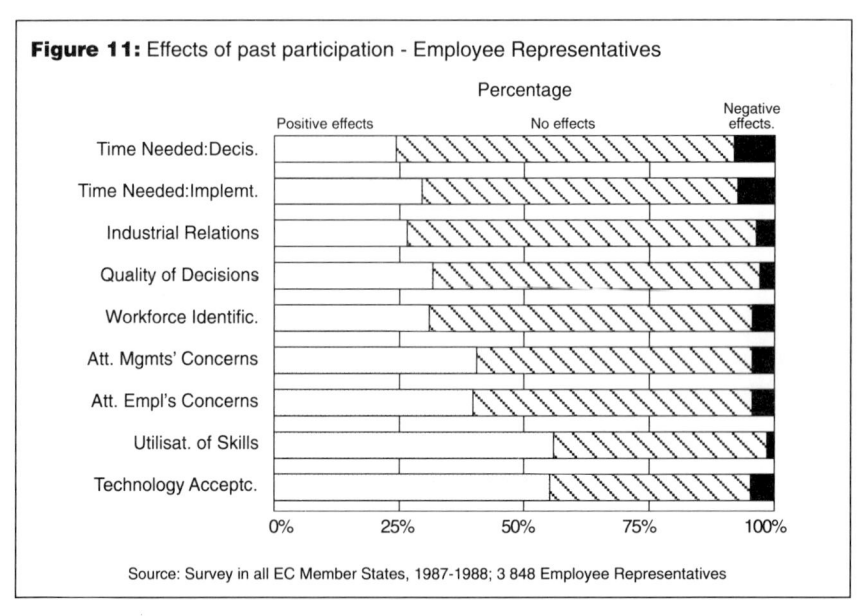

Categories (top to bottom): Time Needed:Decis., Time Needed:Implemt., Industrial Relations, Quality of Decisions, Workforce Identific., Att. Mgmts' Concerns, Att. Empl's Concerns, Utilisat. of Skills, Technology Acceptc.

Legend: Positive effects | No effects | Negative effects.

Source: Survey in all EC Member States, 1987-1988; 3 848 Employee Representatives

1.1 Effects of involvement on decision-making

In general, managers seem to see two problematic repercussions of workforce participation. Both have to do with the **costs of workforce involvement in a time perspective**: 11 per cent of this group of respondents report that participation procedures slowed down managerial decision-making, nine per cent indicate negative effects as to the time needed for technology implementation, another facet of decision-making. These two percentages are not high in absolute terms, but they stand out in relation to the other negative effects reported which vary between one and three per cent only and which can almost be regarded as measurement errors. Low as these shares of retarded decision-making processes and technology implementation may be, we must take this result as an indicator that **participation procedures can sometimes conflict with management's need for quick decisions**.

But we should not exaggerate this danger. First of all, there is the fact that in the majority of cases participation procedures had no effects at all, neither positive nor negative. Then, we have to keep in mind that in many companies workforce involvement had beneficial effects even in regard to the time problem: participation has reduced the time needed to arrive at decisions in 22 per cent of the cases. There are twice as many positive as there are negative assessments. As to time saving in the process of technology implementation, the ratio between positive and negative effects is one to three. Summing up we might conclude that for management **the time aspect is a somewhat sensitive issue. But even so, the benefits of participation clearly outweigh its disadvantages**.

As to decision-making, the overall positive effects are accentuated by the fact that not only the time needed for arriving at decisions has been reduced in almost every fourth case. Every third manager indicates that participation resulted in an increased **quality of decisions**. This result, when the fact that practically no manager voices a deterioration of the quality of decisions, indicates strongly that the somewhat sensitive issue of the time costs of participation might be offset by positive outcomes in another aspect of the issue - namely in the quality dimension.

1.2 Industrial relations and mutual understanding in the firm

Industrial relations generally refers to the relations that exist between management and entrepreneurs on one side and the workforce and their representatives on the other. On a macro-level, the term describes the type of organisation, the written and informal rules and regulations and the type of policies within which the collective organisations of both entrepreneurs and trade unions deal with each other. It covers issues and traditions of antagonistic policies between both parties or a more cooperative attitude.

While the climate of industrial relations in the wider society will clearly be reflected at the level of individual companies, for our methodological purposes we have to treat the individual companies and the information given by managers and employee representatives as company-specific facts alone.

The topic of industrial relations is covered by the direct question as to its development under the impact of participation procedures and through the more indirect question whether the mutual understanding of the parties involved did result in a change of opinion by employee representatives as a result of their involvement. The survey data shows that, from a managerial point of view, participation procedures in technology introduction had no impact on industrial relations in about two thirds of the companies. Negative repercussions are virtually non-existent, but every third manager sees improvements in his relations with his workforce. At first sight, there is a clear trend towards a more favourable industrial relations.

"No impact on industrial relations" is the most common response. This is difficult to interpret. It could mean that these relations were positive to begin with, but it might also be seen as signifying poor industrial relations that did not improve under the impact of participation. With respect to the overall positive impact of past involvement, plus the fact that there is a clear increase of positive assessments with regard to industrial relations, we are inclined to conclude that, **in general, in the companies under study, industrial relations were fairly favourable to begin with, and that the experience of participation in technology introduction has led to further improvements of industrial relations.**

Our interpretation is sustained by the two additional issues which are part of the industrial relations problem: the attention of employees for management's concerns and plans and the attention paid by management to the concerns and interests of employees. Both problems touch upon the social climate within the company as they deal with the question of empathy, i.e. the capability to perceive the problems and interests of the other side.

The impact of participation on mutual understanding within the companies is even more positive than in case of general industrial relations: every second manager indicates that according to his perception both managers and workforces have gained a better understanding of each other's problems as a consequence of participation practices. Again, negative assessments are minimal.

In this context, it is somewhat surprising to find a divergence gap between the shares of reported positive effects as to the development of general industrial relations and the related concepts of mutual understanding. Yet, according to our judgement, this difference is plausible: the nature and quality of industrial relations are developed within long-term processes, and it is unlikely that they are changed in the process of introducing new technology which is of a short-term duration. Under this premise, it is more appropriate to expect that **the improvement of mutual understanding between managers and the workforce will only be one factor among others that will improve industrial relations in the long run**.

1.3 Workforce identification and the concept of company culture

Whereas the time problems of decision-making and technology implementation are rather concrete issues, the industrial relations topic is a blend of the concrete and the "atmospheric": concrete effects of participation might be found in changed rules of formal relations between the parties within companies, whereas the problem of empathy, of mutual understanding of the parties indicates a psychological level which is likely to influence formal regulations. A third set of effect dimensions is basically centred around intangibles which denote the attitude of workers and employees towards their company and towards technological modernisation. Here, we touch upon a broad managerial discussion which gained momentum in the early 1980s: Management became increasingly conscious that a large part of their operating costs are associated with employing people or human resources. The organisational problems which became apparent with the introduction and application of new information technology during the last decade further accentuated the growing dependency of management on the workforces, particularly when dealing with complex technology and qualified workforces.

These new needs are reflected in a new strategy which gained in popularity under the term "human resource management"[2]. The objective of human resource management is to maximise the commitment of employees through the adoption of organic and devolved structures in which the individual employee is encouraged to develop the habits of self-discipline and initiative. More particularly, human resource management occupies a central place in the total "culture" of the firm. This concept of organisational culture gained particular prominence and public attention through the popular success of Peters' and Watermans' "In Search of Excellence"[3] where "successful companies" in the USA are described. Their key ingredients for corporate excellence are a "people oriented culture" and "productivity through people".

The authors view the successful company as a system of common norms, as a **culture** in which the actors contribute voluntarily to achieve shared goals. Although this notion carries some ideological undertones in that it assumes common objectives of all parties involved, it has nevertheless shaped management - workforce relations towards greater empathy of management in regard to workforce interests.

Human resource development and organisational culture are management policies which are geared directly towards employees and workers. They involve complementary forms of participation measures and communication channels like quality circles, project and briefing groups, videos and presentations in addition to traditional union and workforce representation means and channels. In the scope of our study which is concerned with representational participation only, the scope and effects of such direct involvement cannot be assessed from the point of view of employees and workers. But what we are able to find out is - in the research tradition of Daniel[4] for the United Kingdom - is the contribution of indirect, workforce representative participation for the creation of common perspectives between management and the workforces concerning the company's well-being, as reported by managers and employee representatives.

According to the perception of two out of five managers, the identification of the workforce with the company has increased through involving their representatives in the process of technology introduction. This high rate of positive repercussion of participation is even surpassed by two further benefits. As mentioned above, people, their motivation and commitment and their skills are increasingly considered to be a bottleneck in companies which apply modern information technology. Whereas the identification with the company touches upon the commitment dimension, the "utilisation of knowledge and skills of employees" - thus the exact wording of the item in the interview - centres around the heightened need of management to depend on the willingness of the workforces to apply their knowledge and their skills in order to make the riskful endeavour of applying new information technology successfully. In this respect, participation has yielded very benefical repercussions for the companies: 55 per cent of the managers indicate that participation procedures lead to a better utilisation of skills.

The acceptance of technology by employees is considered an important prerequisite for efficient technology introduction and for its proper functioning. People in many western European countries had a poor image of information technology when it made its appearance in the world of work. This technology had the reputation of endangering job security, of intensifying work processes, of making working time more flexible and unregular. Such public apprehensions were obviously not shared by workers and employees in workshops and offices, as Daniel (1987) found out for Great Britain: advanced technical change was very positively received by production workers and office personnel. Such a **basic positive attitude towards new technology is further enhanced when there are opportunities to participate in the introduction and the application of new technology - this is the message of our survey**. Three out of five managers report that workforce representative involvement in the past has led to a heightened acceptance of new technology among workers and employees. Again, negative repercussions are negligible (4 per cent), which means that in roughly one third of the cases only has there been no change of attitude. The positive effects of participation on technology acceptance are distinct and they prevail. **Altogether, there are strong indications that the targets to be achieved through a participatory approach in terms of creating an atmosphere of workforce commitment and cooperation yields exactly those benefits which are expected.**

So far, we have presented the survey results on the effects of participation from the management point of view alone. The **assessment** of the topics discussed above by **workforce representatives conforms basically with the evaluation of their counterparts of the management side**.

When we compare figure 10 and figure 11, the same structure of answers is apparent: as to the majority of issues, workforce representatives see no effects of participation. Negative effects continue to be the exception, and they are clearly superseded by positive outcomes. Although the similarity between both parties is striking, there are two trends which indicate a slight but consistent difference in perception of effects. On average, employee representatives are more sceptical about the impact of participation procedures. This is expressed by higher rates of negative effects in seven out of nine issues and more restricted evaluations of positive outcomes. The cleavage between both parties is largest as to the assessment of the development of mutual concern and understanding. Whilst roughly 50 per cent of the managers indicated positive effects here, more employee representatives are cautious in this respect. Only 40 per cent voice positive changes, and - in comparison to managers - a larger share indicates negative repercussions.

The differences between the parties are not dramatic at all, but the consistency of the differences can be regarded as an indicator for some underlying divergence in perceptions and evaluations. Such a divergence can be expected if we consider the nature of effects which were investigated. In social science literature, in managerial reasoning and in union policies, most of the effect dimensions analysed are regarded to be management concerns first of all. Generally, unions are dealing with more tangible problems of workforce interests. Sometimes they are even opposed to certain ideas like workforce identification with their companies which is often regarded to be a management tool to weaken union-workforce relations. Under these auspices, it is **not the differences** between the two parties **that need attention but the broad agreement between managers and employee representatives** in regard to effect of participation in these issues.[5]

2. Effects of past involvement dependent on intensity and timing of involvement

So far we have given an overall impression on the effects of participation and we have analysed the data without reference to any specific involvement procedures. We now attempt to relate the effects to the intensity of workforce involvement and ask how different intensities (from information to co-determination) produce specific benefits or disadvantages in the companies. This analysis is carried out for selected problems.

We have seen that there were two somewhat critical issues with distinct shares of negative outcomes through participation procedures: the time needed for decision making and for technology implementation. Now, what type of participation could be responsible for delays? Is a situation of weak workforce involvement more conducive towards speedy decisions? Figure 12 holds the data for the issue of decision-making in the planning phase of technology introduction, for both managers and workforce representatives.

When we look at managers first, we learn that co-determination practices slow down the decision making process most of all. Every fifth manager testifies to this point. This is the negative side of the story. On the other hand, there are positive effects under all conditions of workforce involvement, and they surpass the disadvantages of participation in every case. Consultation practices in particular have yielded the most positive results (according to 30 per cent of the managers). But we must be aware that the different types or intensities of involvement do not lead to systematic differences in positive outcomes, and the net gain, the difference between positive and negative effects is generally rather limited.

Employee representatives have a different perspective about the gains and losses generated by involvement. There is a "hard core" of representatives (about 10 per cent) who see decision making to be retarded by all means of participation. As to positive effects, their assessment is more differentiated. First of all, the balance between positive effects - which

Figure 12: Time needed for decision-making dependent on past participation in the planning phase

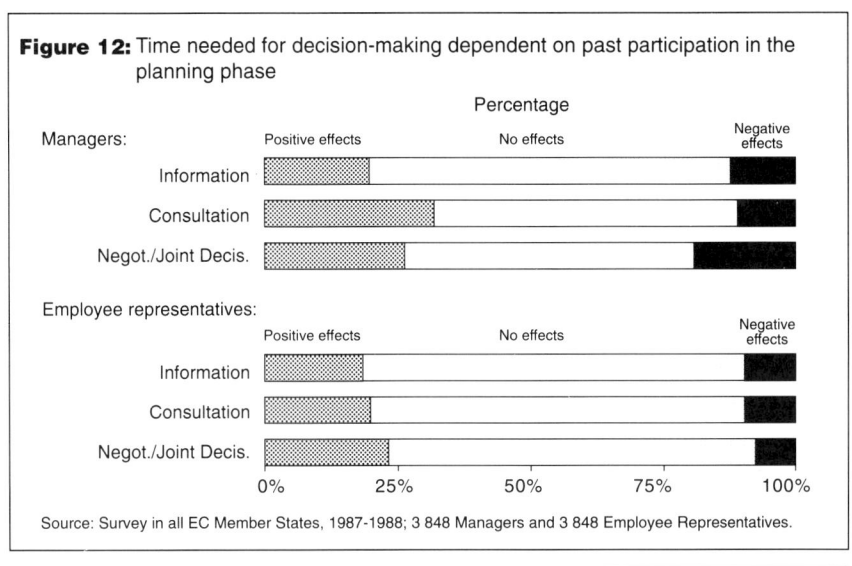

[Bar chart showing Managers and Employee representatives responses across Information, Consultation, and Negot./Joint Decis. categories, with Positive effects, No effects, and Negative effects percentages from 0% to 100%.]

Source: Survey in all EC Member States, 1987-1988; 3 848 Managers and 3 848 Employee Representatives.

Figure 13: Time needed for decision-making dependent on past participation in the implementation phase

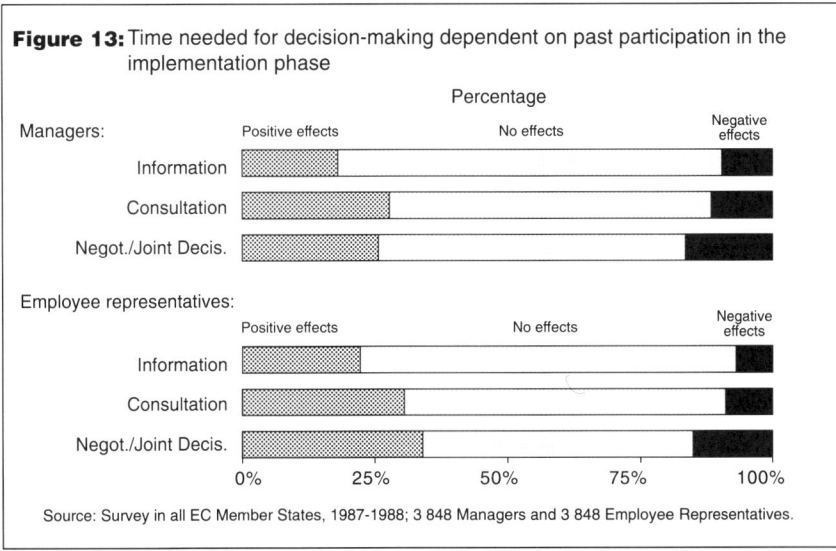

[Bar chart showing Managers and Employee representatives responses across Information, Consultation, and Negot./Joint Decis. categories, with Positive effects, No effects, and Negative effects percentages from 0% to 100%.]

Source: Survey in all EC Member States, 1987-1988; 3 848 Managers and 3 848 Employee Representatives.

means in this case the acceleration of decisions - and the retardation of decisions clearly has a positive slant, and these positive outcomes increase with higher-order, more intense involvement procedures. Thus, one third of workforce representatives thinks that co-determination procedures have speeded decisions up. This result points to the overall picture that workforce representatives see a direct connection between the intensity of their being involved and a positive outcome of participation procedures: the more they are involved the better the result. Managers do not assess the situation in such clear cut and positive terms. According to them, consultation generates the best results.

One result in figure 12 needs particular attention. Whatever means of participation is practiced, in the majority of cases this has no effect on the time problem of decision making: whether workforce representatives were informed or co-determination was practiced, in the majority of companies decision-making was neither slowed down or speeded up. As to this result, managers and employee representatives are in accord. To explain this apparent independence of the time aspect of decision-making from participation procedures poses difficulties which cannot be solved by our data. Instead, this result might point to the possibility that our sample holds quite a few "high trust" companies with well established and smooth industrial relations practices. In such companies, particular participation procedures in regard to specific issues might not play a decisive role, but workforce involvement could be embedded in day-to-day company functioning. As a result, the parties might not be aware of distinct effects of workforce involvement.

In the implementation phase of technology introduction, the different intensities of workforce involvement have basically the same effects on the time aspect of decision-making than in the planning phase. Managers see sizeable time constraints which rise with the intensity of participation. But the positive effects still prevail. Employee representatives see slightly more benefical repercussions than managers. But they have to admit that their involvement creates more time problems in implementing new technology when more intensive participation practices are applied.

In sum, we might conclude that **different forms, intensities of workforce involvement in planning and implementing new technology do not affect the time needed for decision-making in the majority of cases**. Where there are **effects**, they **are more positive than negative**. Generally speaking, consultation and co-determination have the most polarising effects: under these practices the largest gains in the time needed for decision-making as well as the largest time setbacks are reported, from both sides. Obviously, in many companies the involvement of workforces leads to quite different experiences in regard to the time problems in decision-making.

While the time aspect of decision making is coupled with somewhat ambivalent perceptions, workforce involvement leads to a clear-cut picture when it comes to the quality of decisions under the impact of participation procedures. As Figure 14 shows, involvement practices in companies have virtually no negative impact on the quality of decision. If they have repercussions at all, they are positive, and such positive impacts are clearly related to the intensity of workforce representatives participation: the higher the intensity of involvement the better the quality of decisions. Thus, under conditions of co-determination, short of 60 per

Figure 14: The quality of decision-making dependent on past participation in the planning phase

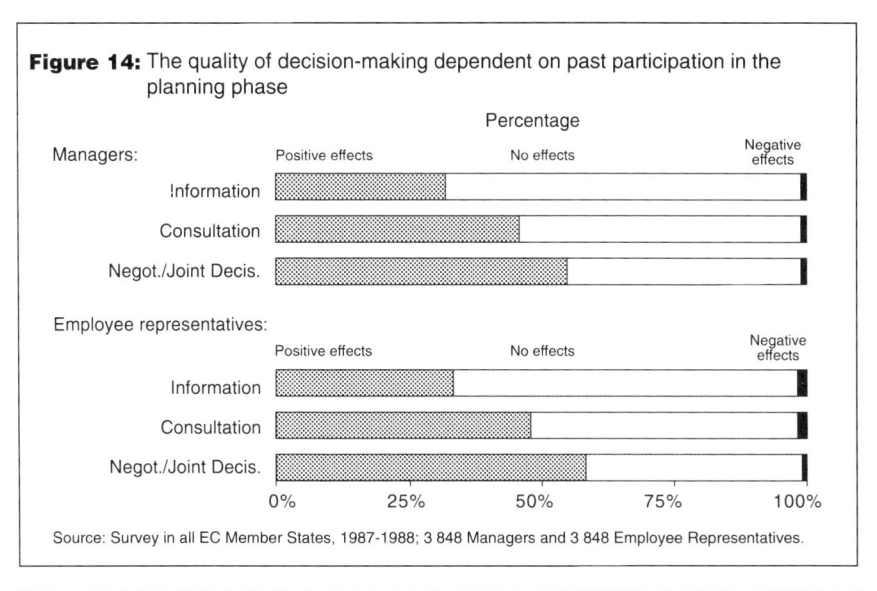

Source: Survey in all EC Member States, 1987-1988; 3 848 Managers and 3 848 Employee Representatives.

Figure 15: Time needed for technology implementation dependent on past participation in the implementation

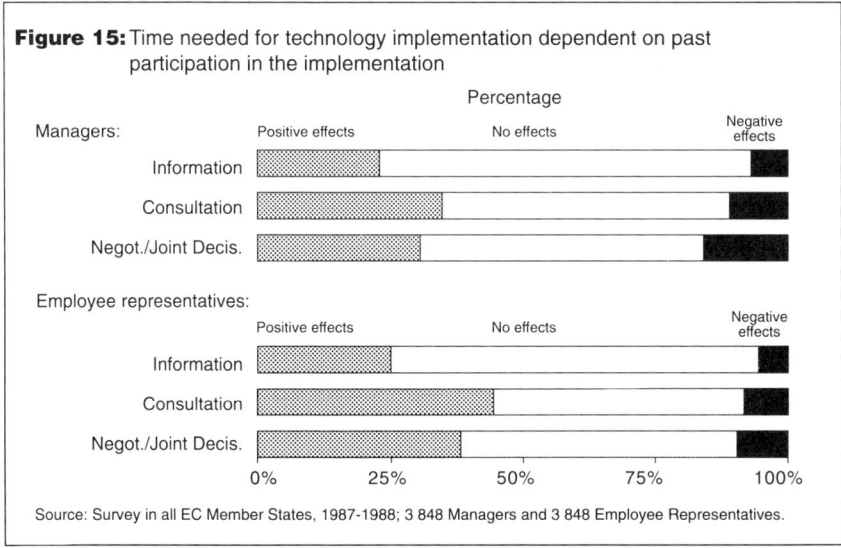

Source: Survey in all EC Member States, 1987-1988; 3 848 Managers and 3 848 Employee Representatives.

cent of both managers and employee representatives indicate a positive impact on the quality of decision-making, whereas in the case of information as the principal means of involving representatives, only about one third of both parties indicate a positive impact. This result does not only apply to the planning phase of technology introduction but the technology implementation as well. From these facts we can safely conclude that **higher intensities of workforce representative participation distinctly improve the quality of decision-making**. And this positive result might counteract the more negative outcome of involvement procedures which have become visible as retardations of decisions. It is conceiveable that the setbacks in one dimension (the time aspect) is counteracted by benefits in another aspect of decision-making, namely the quality of decisions.

Next to the time problems of decision-making, the time aspect in regard to technology implementation was the other issue in which managers in particular voiced relatively distinct criticism (see figures 10 and 11). Although not a large proportion of managers (10 per cent) saw negative repercussions of workforce involvement in the way of slowing down technology implementation, this percentage stands out against the other items where managers pointed out hardly any negative outcomes. In the following section we attempt to investigate in which way the implementation of technology is retarded or supported by workforce involvement.

To exemplify the impact of participation we have chosen the implementation phase[6] as the most appropriate point in time: discussing the time needed for technology implementation can best be done in the case of participation procedures in this phase itself. The data again shows that the different involvement procedures had no effects in the majority of cases. They further indicate that more intense forms of workforce involvement lead to a slow down in technology implementation. Under conditions of co-determination, every fifth manager perceived such a negative outcome. But again, the positive effects prevail under any circumstances which means that whatever procedure was applied, according to 25 to 35 per cent of the managers these procedures were helpful to accelerate technology implementation. In this context, managers had made the best experience with consultation procedures which have been shown to be superior to information giving or co-determination practices. The positive effects of consultation do not deviate very much from other practices. But as employee representatives point to exactly the same fact, there seems to be a systematic value embedded in consultation as a tool that helps to accelerate the application of new information technology, a tool which seems to be superior to other forms of workforce involvement. It is this point at which both parties are in distinct accord. Differences become apparent in so far as employee representatives see, on the whole, the outcome of their activities in slightly more positive terms than managers.

But about every tenth representative does admit that participation did retard technology implementation, regardless what kind of involvement has been in action.

All in all, we find the same pattern of effects as in the case of the time needed for decision-making: there is overall agreement between the parties that workforce participation does not influence the speed of technology implementation in the majority of companies. There is a large number of both managers and representatives who are convinced that the different forms of involvement have helped to accelerate the introduction of technology, consultation practices in particular. There is a smaller percentage on both sides who point to negative repercussions. For them, co-determination practices seem to slow down technology implementation most.

Skills and knowledge of the workforce are potential bottlenecks in the process of technical modernisation of companies. The lack of skills can slow down modernisation processes. Often, the skill level available is an important factor in planning for new technology: complex technologies have a lower chance to be envisaged under conditions of a low average skill level of the company's workforce[7]. In addition, existing skills have to be mobilised to make the introduction of new information technology an undertaking in which the risks are limited. In this context, the involvement of workforce representatives has had a very positive impact.

This became evident in the overview which we presented in figures 10 and 11. But not only the fact of participation plays a role.
The intensity of employee representative involvement contributes positively as well.

For managers, consultation procedures are particularly apt to fully utilise the skill and the knowledge of their workforces to make technology planning a success: almost three out of four managers who have been in consultation processes with their workforce counterparts point to beneficial effects in this respect. According to the managers, all other forms of participation or non-involvement fall way behind the effect of consultation when it comes to make good use of their workforces' skills. Employee representatives express the same positive evaluation, but with a different accent: they see a direct link between the intensity of them having been involved and the utilisation of skill. This means, that according to them skill utilisation improves continually with more intense forms of participation. According to employee representatives, the impacts on skill utilisation are most positive when co-determination procedures, namely negotiation and joint decision-making, were applied. Under such conditions, four out of five representatives point to benefical outcomes of participation[8].

Technology acceptance by the workforce is an important precondition to introduce and to run new information technology effectively. It has been shown (figure 10 and 11) that this acceptance has been high in the companies under study. As figure 17 demonstrates, the high degree of technology acceptance seems to be a structural trait of all workforces[9] which is not too strongly affected by additional involvement practices.

In the planning phase of technology introduction, technology acceptance is high under almost any circumstances. According to both managers and employee representatives, the different forms of participation help to boost such acceptance. But it is comparatively unimportant whether involvement rests on information giving, consultation or co-determination procedures.

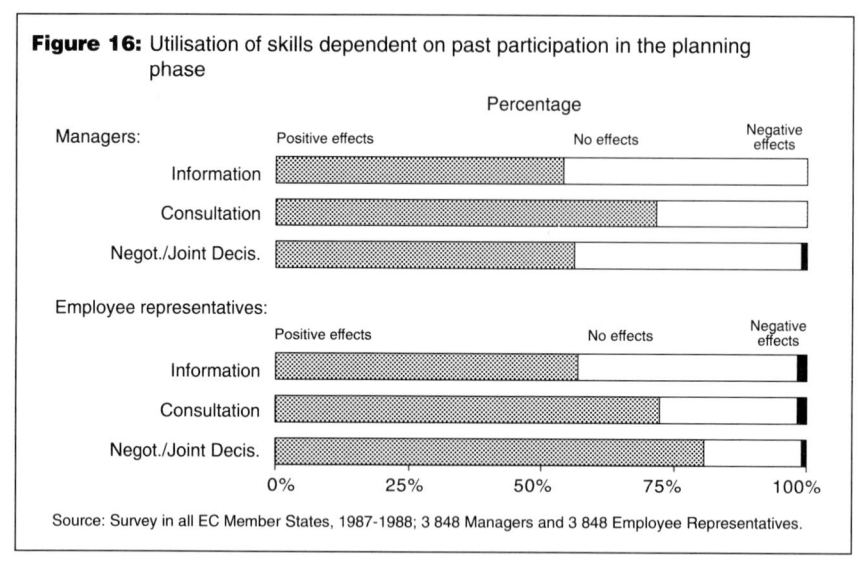

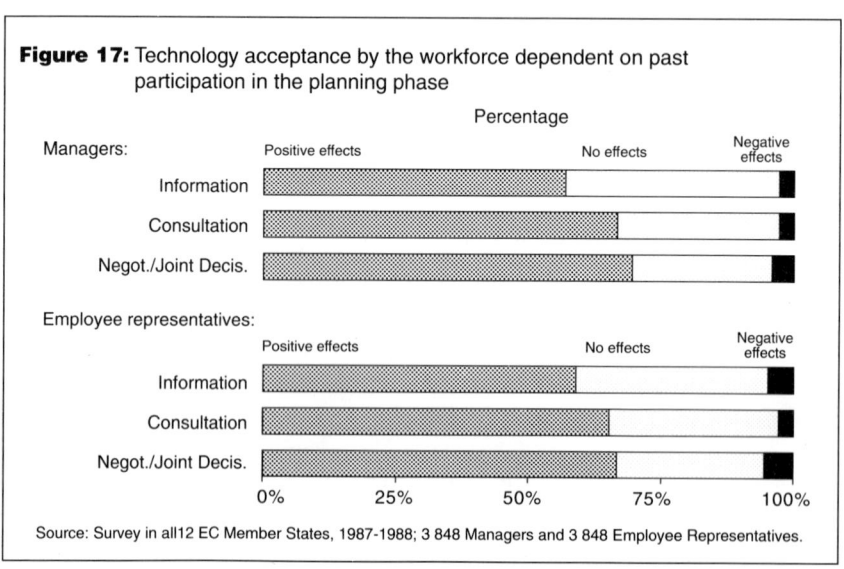

Figure 16: Utilisation of skills dependent on past participation in the planning phase

Source: Survey in all EC Member States, 1987-1988; 3 848 Managers and 3 848 Employee Representatives.

Figure 17: Technology acceptance by the workforce dependent on past participation in the planning phase

Source: Survey in all 12 EC Member States, 1987-1988; 3 848 Managers and 3 848 Employee Representatives.

Although two thirds of both parties indicate that consultation and co-determination have helped to improve the acceptance of technology among the workforces, informing workers and employees about management plans for technological modernisation of the workplaces yielded very positive results as well: almost three out of five managers and workforce representatives point to an increase in technology acceptance due to information provision.

But this relative independence of technology acceptance from participation is mainly a question of the timing. It is valid for the planning phase of technology. But when it comes to the later stage of technology implementation, participation procedures seem to play a more important role, at least in the perception of management.

Our managers tell us that - on the basis of high overall technology acceptance - consultation, negotiation and joint decision-making have particularly positive effects in the process of implementing technology in the workshops and offices. Obviously, they are aware that in this stage of technical modernisation they are particularly dependent on the workforces and that at this point they need direct feedback from the workforces and their representatives. Such feedback is possible in situations of negotiation and co-determination, and these practices obviously support a positive attitude of the workforce towards the technology they have to work with.

Workforce representatives are not as distinct in their positive evaluation of participation effects for the enhancement of technology acceptance. This should not primarily be derived from looking at the shares of positive evaluations which are high and are closely aligned with those of their counterparts on the management side. Here, it is more useful to consider the negative assessments. Though rather small quantitatively, they stand out not only in comparison to the equivalent management perceptions, but in the frame of overall low negative assessments of impacts as well. These negative evaluations do not vary with the intensity of involvement, but remain rather stable regardless of whether there has been no involvement or co-determination. We interpret this result as expressing a general trend to be somewhat apprehensive about the whole issue. We must not forget that technology acceptance is basically a managerial concept which is not on top of workforce representatives's agenda, and many unions in Europe have a very distant attitude towards this concept as well.

The same considerations are valid about workforce identification with their companies. In the concept of organisational culture and corporate identity, measures to boost the identification of workers and employees with the goals and the image of their firms is considered an approach to increase effectiveness. Many unions and union-related workforce representatives are afraid to lose control in case workers and employees identify too much with their companies and with management, thereby neglecting general workforce and union concerns.

At first sight, the data reveals such apprehensions are not justified. In the majority of cases and according to both managers and representatives, the different involvement practices do not affect workforce identification with the company one way or the other. Then, there are large numbers of both parties who point to an increased identification of managers and employees, particularly under conditions of consultation and co-determination procedures. In these instances, managers indicate positive changes up to 50 per cent, and representatives go well beyond 40 per cent of positive impacts. But what gives support to a critical view towards the whole idea of workforce identification with their companies is the fact that a relatively distinct percentage of employee representatives (short of 10 per cent) point to negative developments of workforce identification, regardless of the fact whether they were involved at all or whether they even took part in co-determination procedures. Without overstating our case, we might conclude that as to workforce identification with their companies, participation helps to build up such identification, particularly when it comes to consultation and co-determination procedures. But at the same time there seems to be a hard-core of employee representatives who are rather sceptical about the whole concept. This scepticism probably led them to see basically negative repercussions of participation procedures for the workforces they represent.

3. Conclusions

Reviewing the effects of indirect participation in the technical modernisation of European firms, the overall situation can be characterised as a mixture of no impacts and positive results. When managers and employee representatives indicated effects they were mainly, sometimes overwhelmingly, positive. This is true even for the only sensitive issues of possible time delays through participation: decision making and technology implementation. These are the aspects where managers as well as employee representatives report comparatively high rates of negative effects. But at the same time there are always more respondents who indicate that involvement procedures have helped to accelerate decision-making and technology implementation. In addition, there are comparatively high rates of both managers and workforce representatives who indicate an increased quality in decisions, and practically nobody voices negative repercussions in this respect.

Figure 18: Technology acceptance by the workforce dependent on past participation in the implementation phase

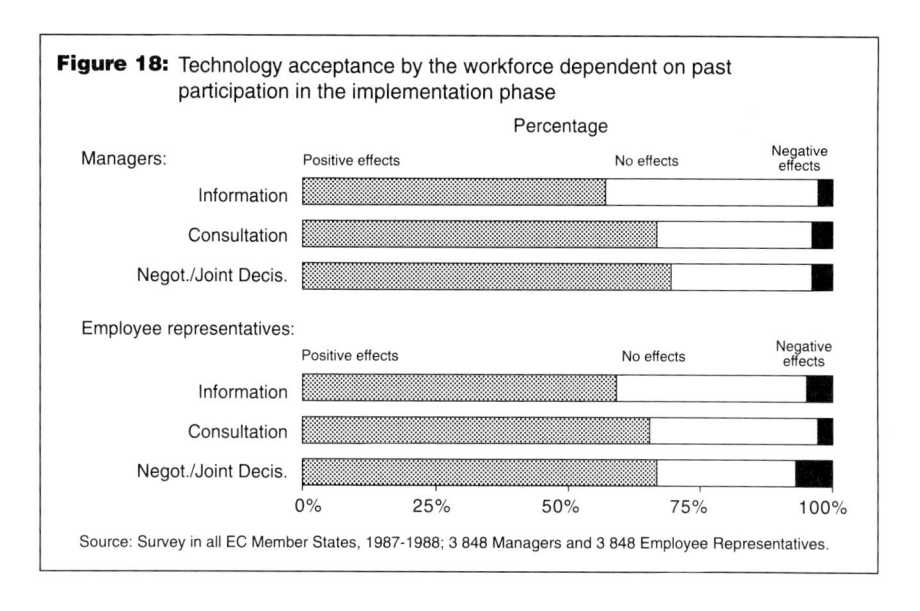

[Bar chart showing Managers and Employee representatives responses for Information, Consultation, and Negot./Joint Decis. categories, with Positive effects, No effects, and Negative effects percentages]

Source: Survey in all EC Member States, 1987-1988; 3 848 Managers and 3 848 Employee Representatives.

Figure 19: Workforce identification with the company dependent on past participation in the implementation phase

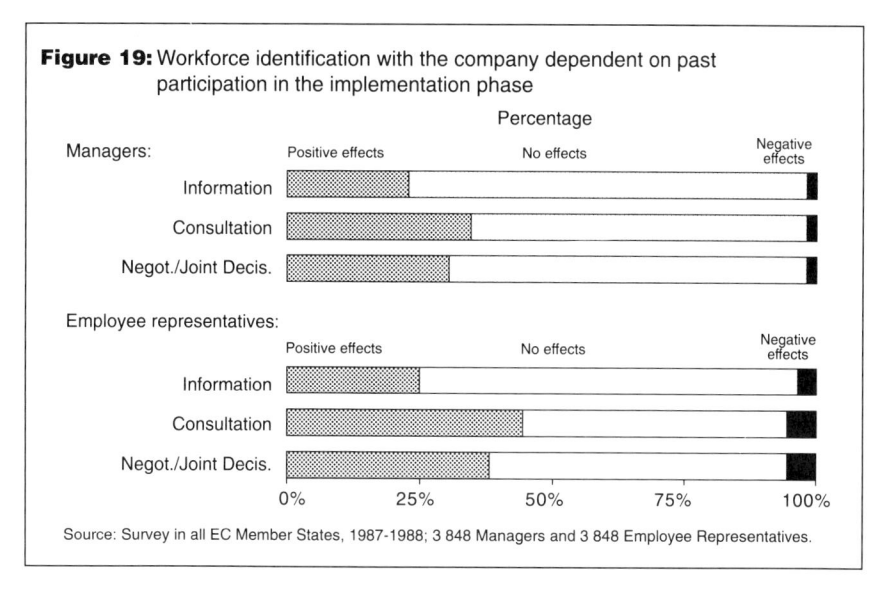

[Bar chart showing Managers and Employee representatives responses for Information, Consultation, and Negot./Joint Decis. categories, with Positive effects, No effects, and Negative effects percentages]

Source: Survey in all EC Member States, 1987-1988; 3 848 Managers and 3 848 Employee Representatives.

Further, industrial relations and the mutual understanding of both management and the workforces for each others' problems and constraints have been improved under the impact of participation in many companies. We have reason to believe that these positive developments are embedded in a context of predominantly high trust relationships which existed prior to the experience of participation in technical change, but which have since forged a more improved co-operative climate. The majority of both sides indicate an increase of technology acceptance and the utilisation of workforce skills due to participation. Knowing that new technology is generally well accepted by the workforces and that new technology is best applied under conditions of a highly skilled workforce, these results, again, point out to conditions in the companies which were conducive to participation to begin with.

What kind of participation procedures and which intensity yields the best results? As to the time needed for decision making, information giving, consultation or co-determination procedures do not discriminate much as to their effects. As a trend, in management's view it is consultation practices which have had the strongest positive impact while co-determination procedures carried the strongest potential to slow down decision-making processes. Employee representatives have a slightly different perception in that they attach the highest positive influence to negotiation and joint decision-making practices. But the differences between both parties are far from dramatic. In case of technology implementation, both sides are in accord that consultation practices are most conducive to accelerating decisions. As to other dimensions, there is a general common understanding that positive outcomes of indirect employee involvement increase with more intense form of participation. Only when it comes to the utilisation of workforce skills, are managers distinctly in favour of consultation procedures whereas employee representatives evaluate negotiation and joint decision procedures at the most effective.

But the differences between management and employee representatives in regard to the overall assessment of benefits and disadvantages of participation as well as in their perceptions of the kind of participation procedures which have generated effects are rather small. As a general rule one might say that workforce representatives are somewhat more hesitant to point out to positive impacts than managers, and there seems to be a small hard core of representatives who indicate negative effects throughout. But again, these differences are minor on the whole. Such a result is surprising in view of the fact that most of the effect dimensions can be summarised as management concepts. These are issues which are predominantly discussed in management literature and which centre around the idea of creating a corporate identity, creating emotional ties between the company and the workforce and to foster technology acceptance in order to improve productivity of the enterprise. Under such conditions, it is not the slight discrepancies but the overall high agreement between both sides which needs attention. This basic consent which we encountered already when discussing the facts of past participation leads us to conclude that we are dealing with largely co-operative, high-trust relationships between both sides in the European companies under study. Both parties convey the message of rather high consensus between them. Such consensus carries the view of participation to work as a productive force to achieve common ends. The situation does certainly not point towards antagonistic relationships between management, the workforces and their representatives. Yet, the overall consensus does not give any clues in what way both parties want to move on to a situation of industrial democracy in the future. It might as well be that both sides are content with the situation they are operating in at present.

NOTES

1. The repondents could voice further positive or negative effects. This open question did not elicit much additional information.

2. Keith Sisson (ed.): Personnel Management in Britain, London 1989; OCED 1988, op. cit.

3. T. J. Peters and R. H. Waterman: In Search of Excellence, New York 1982.

4. Daniel, op. cit.

5. Such an overall agreement as to the outcome of participation points to the methodological problem of isolating cause and effect. Our selection of companies might have been a positive one in the way that high trust companies with cooperative management-union relations are over-represented in our sample. When such parties engage in participation procedures there is a likelihood of positive results, simply because of the basic advance agreement between the actors in many action fields, technology introduction included. The overall small shares of negative effects of participation might be a case in point. Further, under conditions of basic trust and cooperation, participation procedures have a high liklihood of even ameliorating the positive situation. The many positive outcomes of participation in our survey might substantiate this interpretation. Therefore, it would be premature to interpret the positive changes in attitudes as a result of participation practices only. Although we have no direct indications, we are well advised not to exclude the possibility that the overall positive picture of improvements within the companies under study is not due to participation practices alone. Developed organisational cultures and cooperative industrial relationships prior to participation procedures in technology introduction might also have made such outcomes possible.

6. Data analysis shows basically the same results as to participation in the planning phase, the only differences being slightly larger shares of negative outcomes and slightly less positive changes under all participation procedures in the planning phase.

7. Sorge A. et al.: *Microelectronics and Manpower in Manufacturing: Applications of CNC in Great Britain and West Germany,* London 1983.

8. Again, the same pattern of response is valid for the phase of technology implementation, with slightly smaller shares of positive results, but without any increase of negative impacts.

9. This is, in fact, one of the central messages of DANIEL'S research on Great Britain as well. See DANIEL, op. cit.

Chapter Six

Future Participation

1. Introduction

A review of our research findings so far reveals an ambivalent situation of participation in technology introduction: As to the intensity and phasing of involvement, the intensity was low, particularly in the strategic phase of technology planning. It was somewhat higher in the operational phase of technology implementation. As to content matters, workforce involvement in issues of primary interest to management turned out to very restricted on the whole. In many cases there was no involvement whatsoever. More positive from a workforce representative point of view are the participation opportunities in issues which, by tradition, have always been the domain of unions and workforce representatives. In matters of work organisation, task specification, job satisfaction, and health and safety, workforce involvement has been rather prominent, not only as information and consultation but also as bargaining procedures.

Among the means of participation, the soft forms of information and consultation predominate in all instances. This meets the minimum standards both social parties agreed upon in Val Duchesse. Half of the workforce representatives interviewed assessed the quality of information they received in outright positive terms. Only a few reported outright negative experiences.

So far, the experience of participation in the past can best be characterised as one of ambivalence - positive and negative aspects exist side by side. When we look at the **effects** of past workforce involvement, the situation reveals distinctly more positive aspects. Both parties interviewed testified to the point that workforce participation hardly resulted in any disadvantages which complicated the proper functioning of companies. In issues where both parties identified effects of participation at all, the positive evaluation of these effects was almost unanimous. Such positive repercussions were particularly frequent in matters that centre around social relations and common understanding between the workforces and management. In many cases participation has helped to strengthen corporate identity, prepared the ground for making full use of workforce skills and helped the workforces to accept the new technologies in their companies.

At Val Duchesse, both employers and unions stated that "information and consultation must facilitate and should not impede the introduction of new technology...". The practice of participation procedures analysed so far fully supports these expectations of the parties concerned. According to large numbers of respondents from both sides participation procedures have proven their capability to facilitate the process of technological modernisation; impediments to such a process were rarely voiced. Thus, apprehensions about workforce involvement to endanger a smooth and flexible technology introduction have been shown to be premature.

Considering the intentions of both parties as to the future of participation three forms of attitudes are possible in theory:

1. The parties want to reduce workforce involvement.
2. They want to maintain the level and the means of participation procedures which they have excercised so far, and
3. they want to develop such instruments further and increase the overall level of worforce involvement.

According to the overall positive effects both parties experienced in the past, the first possibility is very unlikely. There is no real reason to expect neither managers nor workforce representatives to aim at participation procedures in the future which go beneath a practice which has proven its worth to either of the parties. Under the conditions we met so far we are more likely to see a pattern of future participation which can be characterised as a mixture of continuity and change towards an increase of workforce involvement. Both forms of reactions are possible under conditions of positive experiences in the past: a successful policy can be maintained; the parties involved do not see any need to change it. Then, because of the proven worth of a policy, those successful measures might be developed even further; participation has been shown to serve the needs of either party so well that it would only be logical to extend its future use, thus producing even increased benefits.

The possible attitudes and expectations of both sides as to the future of participation have been discussed in the frame of reference of managers and workforce representatives acting in the enterprises and workshops. A further interesting question arises in the context of the organisations of both entrepreneurs and unions: how do the future intentions of the "rank and file" of UNICE and ETUC compare with the strategies pursued by their organisations? Are they in accord or do the members disagree with their organisation?

For the entrepreneurial side, no problem is to be expected. The UNICE position can be characterised as one of voluntarism. UNICE's stand is that their members are free to choose any measure they like. If managers think it appropriate to go into co-determination procedures they may do so. If they do not intend to involve workforce representatives at all and if legal and other circumstances permit, they may choose this strategy as well. Such management's behaviour is always in accord with the overall goals of UNICE. The situation is somewhat different for employee representatives. ETUC, the European Trade Union Confederation, has adopted a model of industrial democracy. In this strategy, they want their organisation and their representatives in the companies to be involved in all company matters and as early as possible, and as to the intensity of involvement they favour co-determination procedures. The question arises here as to whether the union activists in the companies and at the shop floor level are in accord with such a strategy or whether they have different opinions about workforce representation and favour other less encompassing and demanding approaches. Thus, the data are apt to describe expectations and aspirations of workforce representatives, of the "rank and file" vis à vis their organisations'.

2. Intensity and timing of future participation - the aggregate level

It is this frame of expectations in which the intentions of both managers and employee representatives as to future participation are analysed. These intentions are presented on the background of their past experiences in their companies. We first look at the managers and their attitudes as to workforce participation in the planning and the implementation phases of technology introduction.

The overall result from figures 20 and 21 is: **Managers intend to increase participation and want to develop more intense forms of workforce participation in the future.** In detail: Looking at the situation in the planning phase, one of the most striking results is that "no involvement" which was reported by two out of five managers in the past has lost lots of its attraction for the future. Only one out of five management respondents (18 percent) wants to keep workforce representatives completely out of future planning of technology introduction. This reduction of 21 percentage points in case of "no involvement" in the planning phase does not lead to an increase in more future information provision. The number of managers which want to practice information as the only means to involve employee representatives is virtually the same for the past and for the future (39 and 38 percent). But **consultation procedures have gained particular attraction**: 12 percent of the managers practiced it in the past, and 27 percent want to go into consultation in the future. A similar, not quite as distinct change in intentions is to be seen as to negotiation and joint decision making. Such binding procedures were rather rare in the past (10 percent). But 17 percent of the managers regard it an attractive means of future participation.

As to the **implementation phase**, the overall pattern resembles that of the planning phase. But the level, the intensities of intended workforce involvement are higher. "No involvement" in the implementation phase was comparatively low in the past. For the future its share has decreased

Figure 20: Comparison of past and future involvement in the planning phase - Managers

Percentage

- No Involvement: Past 39, Future 18
- Information Only: Past 39, Future 38
- Consultation: Past 12, Future 27
- Negot. Joint Decis.: Past 10, Future 17

Intensity of Involvement

Source: Survey in all EC Member States, 1987-1988, 3 848 Managers

Figure 21: Comparison of past and future involvement in the implementation phase - Managers

Percentage

- No Involvement: Past 22, Future 10
- Information Only: Past 39, Future 30
- Consultation: Past 21, Future 31
- Negot. Joint Decis.: Past 18, Future 29

Intensity of Involvement

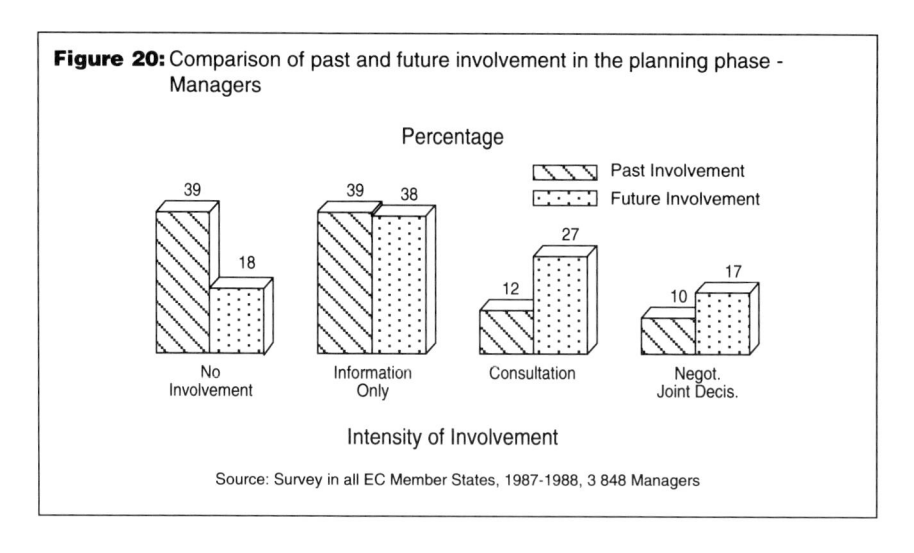

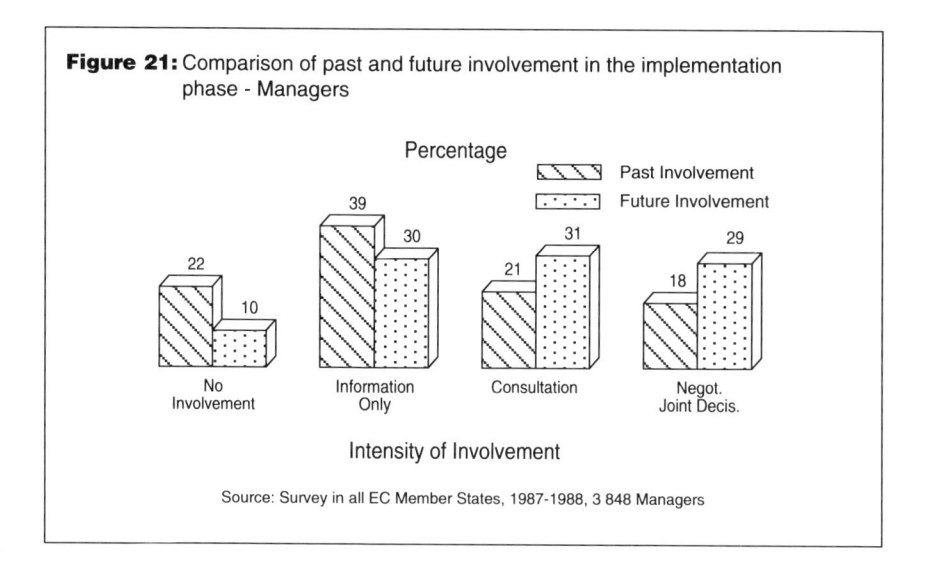

Source: Survey in all EC Member States, 1987-1988, 3 848 Managers

again by more than half, and only one out of ten managers does not want to involve employee representatives in this operational phase of technology introduction. Even information provision as the only participation means has decreased by a fourth (to 30 percent). At the same time, consultation, negotiation and joint decision making have become attractive instruments to support the process of technology implementation in companies. **Three out of ten managers want to consult the other side in the future, and an equal share envisages bargaining procedures of negotiation and joint decision making.**

As to both phases, the change of management's attitudes towards participation is very distinct and outright positive for the issue of workforce involvement. In no case are weaker forms of participation intended in the future as compared with the past. But this observation applies to groups of respondents only. It does not allow hints as to individual changes of attitudes and intentions. Thus, when looking at the planning phase (figure 20), for instance, we are not able to interpret a 21 percentage point reduction as to "no involvement" and a 22 percent increase in favour of consultation, negotiation and joint decision making in a way that intentions have shifted right from "no involvement" attitudes towards the higher forms of participation. Our aggregate figures do not permit us to assess individual attitudinal changes, even changes from higher forms of involvement towards information provisions or even no involvement. To trace such developments we have to analyse the responses on an individual level of comparison (which will be done in the following paragraph). But what we can deduce from our aggregate data is that downward changes in attitudes among managers cannot be too frequent. The constant increases toward more intense forms of participation make such negative changes very unlikely.

Comparing management attitudes with the intentions of employee representatives yields a picture which can be characterised as agreement and dissent: like management, **employee representatives** want more involvement in the future - as to this point, both sides are in accord. But they disagree as to the extent of future involvement: workforce representatives **aim at much higher intensity of participation as compared to what managers envisage.**

For the planning phase, 43 percent of the workforce respondents indicated "no involvement" in the past. Obviously, this situation was evaluated very negatively: only 9 percent of the representatives vote for the continuance of this situation. According to this result we might conclude that on the European level only one out of ten workforce representatives does not want to be occupied at all with basic decisions concerning technological change. The wish to be involved even in fundamental, strategic matters of technological modernisation is evident among European workforce representatives. As to the intensities of involvement, the aspirations are almost evenly distributed among all three forms, with a slight tendency towards gaining genuine influence. Only just under a quarter of workforce representatives are content with being informed, making information provision not too popular a concept. Instead, they favour procedures of consultation and, first of all, negotiation and joint decision making. As to both policy means, the **desire to be drawn into future technology planning is almost three times**

Figure 22: Comparison of past and future involvement in the planning phase - Employee representatives

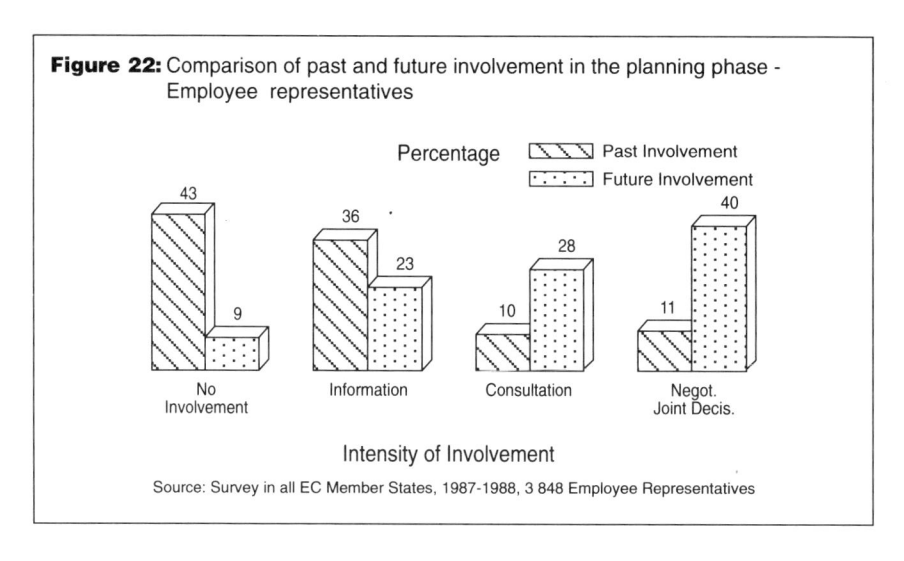

Percentage — Past Involvement / Future Involvement

- No Involvement: 43 / 9
- Information: 36 / 23
- Consultation: 10 / 28
- Negot. Joint Decis.: 11 / 40

Intensity of Involvement

Source: Survey in all EC Member States, 1987-1988, 3 848 Employee Representatives

Figure 23: Comparison of past and future involvement in the implementation phase - Employee representatives

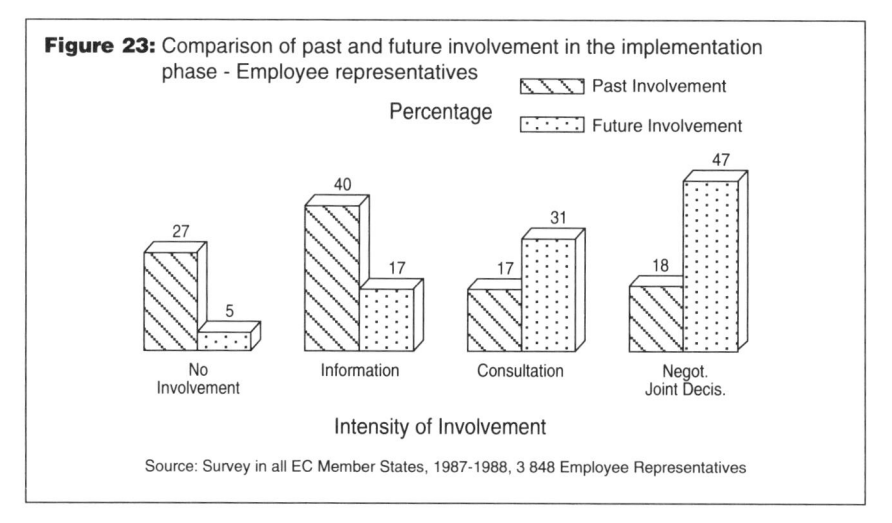

Percentage — Past Involvement / Future Involvement

- No Involvement: 27 / 5
- Information: 40 / 17
- Consultation: 17 / 31
- Negot. Joint Decis.: 18 / 47

Intensity of Involvement

Source: Survey in all EC Member States, 1987-1988, 3 848 Employee Representatives

as strong as the chances they had for such participation in the past: 28 and 40 percent want to be consulted and envisage bargaining chances in the future.

This trend for high level participation provisions among work-force representatives becomes particularly distinct in the **implementation phase**, the traditional field of workforce interests. Not to be involved at all which was reported by every fourth respondent for the past has become almost a residual category (5 percent). It is evident that in their traditional action fields employee representatives do not want to be shut off from participation. Only 17 percent is content with only being informed about problems of technology implementation. They clearly aim at high level intensities of involvement even beyond consultation:

almost every second European representative (47 percent) **wants to have the opportunity to negotiate and co-determine the practical repercussions of technical change in their companies.**

Comparing the **aspirations of both parties** for the future, we can state a **considerable overlap of intentions as well as distinctly differing views**. Both European managers and workforce representatives want to decrease the tradition of no participation substantially. With one exception - managers in the planning phase - they even want to devaluate the policy of information provision in favour of more sophisticated means of workforce involvement such as consultation, negotiation and joint decision making. In regard to the binding participation forms of the bargaining type, sizeable numbers of management are ready to go beyond the standards laid down in the Val Duchesse Agreement, namely information and consultation. As to this increase in bargaining type of procedures both managers and workforce representatives are in high accord.

Yet, the empirical results uncover a certain potential of cleavage and possible political conflict as well. **As to negotiation and co-determination procedures, the aspirations of European employee representatives distinctly go beyond those of European management intentions**: In the planning phase 17 percent of the managers and 40 percent of their counterparts are aiming at binding participation procedures, and in the phase of technology implementation the respective shares are equally far apart (29 and 47 percent). A similar problematic cleavage we find in the problem of involving workforces at all. Here, too, the opinions of both parties differ considerably. It is exactly at these extremes of the continuum of participation procedures where the future problems are located. For quite a few European companies this might mean a situation of dissonance between the social parties involved. On the other hand, we must not disregard the considerable common understanding both European managers and employee representatives have already reached as to common procedures and co-operation in the process of technological modernisation of their firms. All in all, it appears that the common ground for cooperation reached already in the European Community is much wider and more solid than the pitfalls which undoubtedly exist.

There is a further, more indirect level of **comparison** which refers to the intentions of workforce representatives as the actual policy makers and the strategies of unions about desirable participation systems and solutions. We have already pointed out that the European Trade Union Confederation (ETUC) favours co-determination procedures at all levels of company decisions. How does this political goal relate to what the rank and file of unions envisages for the workforces they represent? For such a comparison, we must be aware that only half of the workforce respondents who answered the question about their union affiliation are union members. In a strict sense, a comparison between the ideas of workforce representatives and European trade union goals must be limited to union members. Non-union representatives might vote differently, being neither emotionally nor organisationally tied to union policies. In our data, this possible split in representatives' attitudes did not really appear.[2] Europe wide, the **actors at company and workshop level** have basically the same ideas about future participation, regardless of union membership. And **their attitudes are not really in accord with the ETUC position**: as to the strategic planning phase of technology introduction, 40 percent of workforce representatives vote for co-determination practices in the future. But information and consultation are preferred by every second practitioner of participation. Even for the operational phase of technology implementation where workforce interests are more directly affected, **co-determination procedures are not the first option of employee representatives**: the preferences for co-determination procedures and information and consultation are equally distributed. For the unions and for ETUC these results indicate that a participation strategy in the scope of industrial democracy with far-reaching union and employee rights, does not meet the intentions of a large part of the union rank and file in the European Community.

3. The dynamics of participation - the individual level of comparison

The comparison between participation levels in the past and intended participation in the future has revealed a distinct trend for more workforce involvement in the future. With differing degrees, this applies to managers and employee representatives as well. We arrived at these results by comparing group percentages for the diverse participation procedures, separate for the phases and the type or respondent. Such an approach gives an overall picture, but has three shortcomings which restrict insights and interpretation:

1. It does not allow us to **trace individual changes** in attitudes towards workforce participation. For instance, 39 percent of the managers indicated that involvement in the planning phase took place in the form of information. An equal share of managers (38 percent) wants to excercise information provision in the future (see figure 20). A common sense interpretation would be that these managers did not change their minds but continue to practice an established pattern of involvement. But both groups of managers might be quite

differently composed: as to future intentions, some respondents might have practiced no involvement in the past and might prefer information provision. Others might have had experience in consultation procedures but prefer to revert to information practices in the future. Those who did practice information might now prefer no information or consultation etc.

2. Aggregate level analysis does not allow any assessment of **how far-reaching the change of attitudes** has been. Do attitudinal changes occur over just one intensity of involvment, say from no involvment to information provision, or are there distinct changes over two or three intensities?

3. Finally, aggregate level analysis gives no clues as to the **direction of attitude change**: Do respondents with high level experience in participation procedures revert to lower levels under the impression of this experience or is there a general trend to move on to higher involvement levels? And how far reaching are such trends?

In sum, the method applied so far does not allow for the precise **analysis of attitudinal stability and change**[3]. We arrive at such information by contrasting individual responses as to the past with the aspirations the same individuals have about future participation. Figure 24 holds this information for the phase of technology introduction from the management point of view.

In figure 24, the very left part of the graphical presentation holds all management respondents who had indicated "no involvement" in their companies at the time of interviewing or in the past. This group is now differentiated according to their views and intentions about participation in the future. The data reveals that two out of five managers (43 percent) did not alter their attitude towards the involvement of workforce representatives: they do not intend to inform them about technology introduction in the planning phase, let alone consult them or go into bargaining processes. But the remaining 57 percent have altered their ideas about the participation issue: 27 percent want to inform employee representatives in the future, another 21 percent go even further and envisage consultation procedures. And 9 percent indicate a radical change of practice in that they want to negotiate and bargain about the planning of technology introduction with their workforce counterparts.

Examining figure 24 in this frame of reference, we can state that **stability of attitudes is the most common feature among managers**. As to the planning phase, every second manager who practiced information provision in the past intends to go on with this policy. Three out of four managers want to go on with consultation practices which they applied in the past, and 84 percent of the management's repondents who have

Figure 24: Dynamics between past and future involvement in the planning phase - Managers

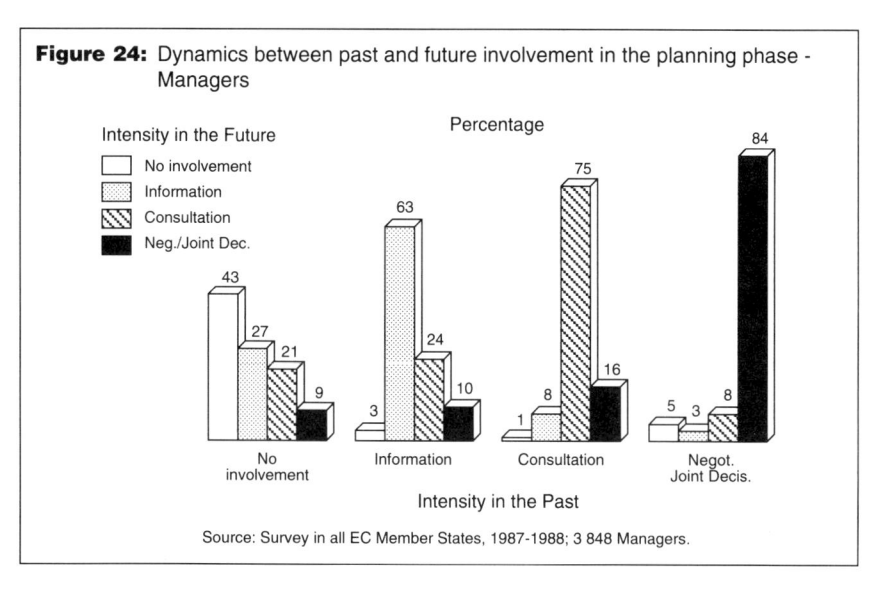

Intensity in the Future:
- No involvement
- Information
- Consultation
- Neg./Joint Dec.

Intensity in the Past (No involvement): 43, 27, 21, 9
Intensity in the Past (Information): 3, 63, 24, 10
Intensity in the Past (Consultation): 1, 8, 75, 16
Intensity in the Past (Negot. Joint Decis.): 5, 3, 8, 84

Source: Survey in all EC Member States, 1987-1988; 3 848 Managers.

Figure 25: Dynamics between past and future involvement in the implementation phase - Managers

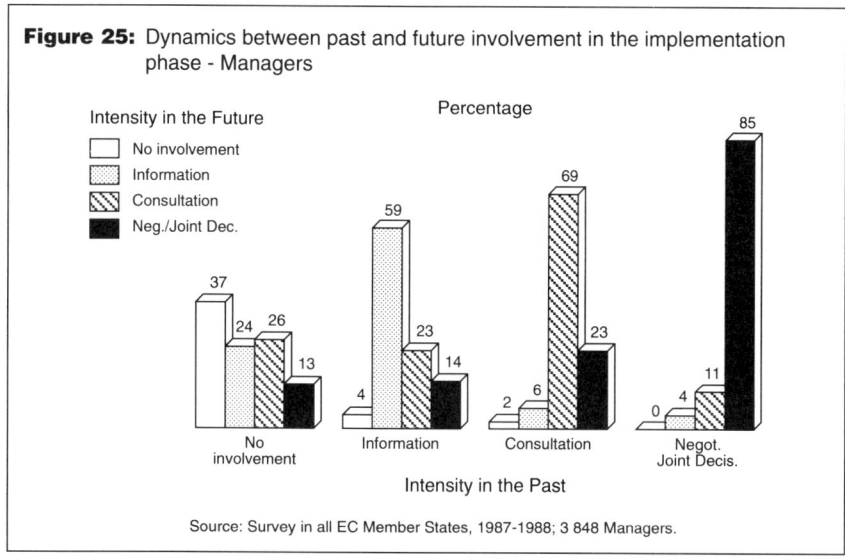

Intensity in the Future:
- No involvement
- Information
- Consultation
- Neg./Joint Dec.

Intensity in the Past (No involvement): 37, 24, 26, 13
Intensity in the Past (Information): 4, 59, 23, 14
Intensity in the Past (Consultation): 2, 6, 69, 23
Intensity in the Past (Negot. Joint Decis.): 0, 4, 11, 85

Source: Survey in all EC Member States, 1987-1988; 3 848 Managers.

been in situations of bargaining and negotiating technology introduction intend to apply such practices in the future, too. Very decisive for the topic of participation is the fact of a clear relationship between the intensity of involvement and the stability of attitude: The lower the intensity of past involvement the higher the rate of attitude change towards more intense forms of workforce participation. But: **the higher the intensity of past participation the higher the rate of managers who want to carry such participation procedures into the future**. Thus, among European managers, the experience of higher level participation has a decidedly higher likelihood of being maintained than low level workforce involvement in the past. Quite obviously, management's experience with workforce involvement in technology planning has largely been felt to be rewarding and they intend to maintain such policies in the future.

Although stable attitudes characterise the situation of managers in regard to participation in the phase of technology planning, there are attitude changes. They vary in scope and direction. As to the direction of attitudinal change, there are logical limits at the two extremes of the continuum of participation procedures. Where workforce representatives were not involved in the past, any changes can occur only towards more and more intense forms of participation. In case of negotiation and bargaining procedures which were applied in the past, change, by necessity, can only be conceived as refraining from such measures in the future. As to the latter procedures, 14 percent of the European managers interviewed intend to avoid such binding means of participation in the future. There is no clear cut pattern as to how far they want to restrict workforce representative involvement: 8 percent intend to revert to consultation procedures, 3 percent to information provision, and another 5 percent want to cancel the whole idea.

At the other extreme, there are considerable changes of attitude towards future participation in technology planning. All these changes mean that, for the future, **managers intend to move out of a situation of no workforce involvement**: Every fourth manager now votes for information provision, every fifth manager wants to consult workforce representatives, and one out of ten management respondents even goes as far as going into negotiation and bargaining procedures. All in all, a large share of managers (57 percent) who did not practice any involvement in technology planning want to alter this situation, partly with far reaching aims as to the intensity of future workforce participation.

Such attitude changes towards higher intensities are dominant, too, in the case of managers which so far had supplied information only or have been in consulation procedures with workforce representatives, even to the point of reaching negotiation and bargaining arrangements. This forward thrust is particularly visible where managers used to inform about technology planning in the past. For the future, 24 percent want to go into consultation situations, and another 10 percent envisage negotiation and joint decision making procedures. Where consultation was the dominant participation procedure, only 8 percent of the managers wish to revert to information provision, but twice as many aim at a bargaining situation in the future.

As to future involvement in the phase of **technology implementation**, this pattern of stability and change of management's intention **virtually**

resembles the pattern of the planning phase. Stable, unchanged attitudes dominate, and where there are changes they point to higher intensities of workforce participation. Here, the trend to move into situations of negotiation and joint decision making are more pronounced than in the planning phase. Management's increased readiness for bargaining procedures are easily understood as - from a management point of view - technology implementation is of secondary importance, but it strongly needs the cooperation and a positive motivation of workers and employees to make the new technology function.

Reviewing the dynamics of management's attitudes towards the idea and practice of involving workforce representatives in the process of technological innovation, three results stand out:

- The most frequent attitude is that of stability: what has been excercised in the past is intended to be carried over into the future.

- Attitudinal stability is highest in cases of negotiation and joint decision making procedures, and it decreases continually towards the situation of no workforce involvement. Here, the share of managers who have altered their intentions as to future participation is highest.

- Where attitudes have changed these changes point distinctly towards procedures of higher intensity. Managers who want to reduce future participation to levels lower than exercised in the past are the minority.

European managements's position about the future of participation in processes of technological modernisation can **best be described as a combination of retaining practices applied so far and of carefully moving on towards extension and intensification of involvement procedures**. This attitudinal pattern is in rather **distinct contrast to the ideas employee representatives** have about future involvement. Stability was the dominant pattern of management attitudes - workforce representatives' intentions can best be characterised as change: **change towards participation procedures of consultation, negotiation and joint decision making**.

Figures 26 and 27 testify to the point. For employee representatives, we find stability of past participation experience and future intentions only in cases of higher level participation procedures. In case of negotiation and joint decision making, nine out of ten workforce representatives want to maintain such procedures in the future. This applies to both the planning and the implementation phase, and this attitude is largely in accord with management's ideas: both parties which have been dealing with each other on this highest level of involvement want to carry on such experience in the future. As to consultation procedures, every second European workforce representative wants to maintain such a relationship in the planning phase of technology introduction. But a high proportion (39 percent) would like to go into processes of negotiation and joint decision making - more than twice as much as managers are ready to accept.

In both the planning and the implementation phase, not to be involved is highly unpopular among the workforce respondents. Only 14 and 18

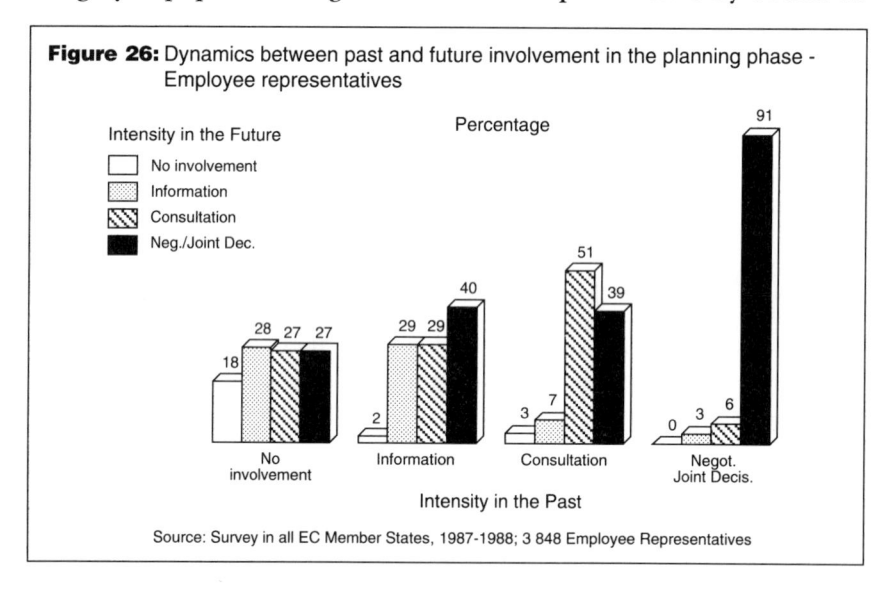

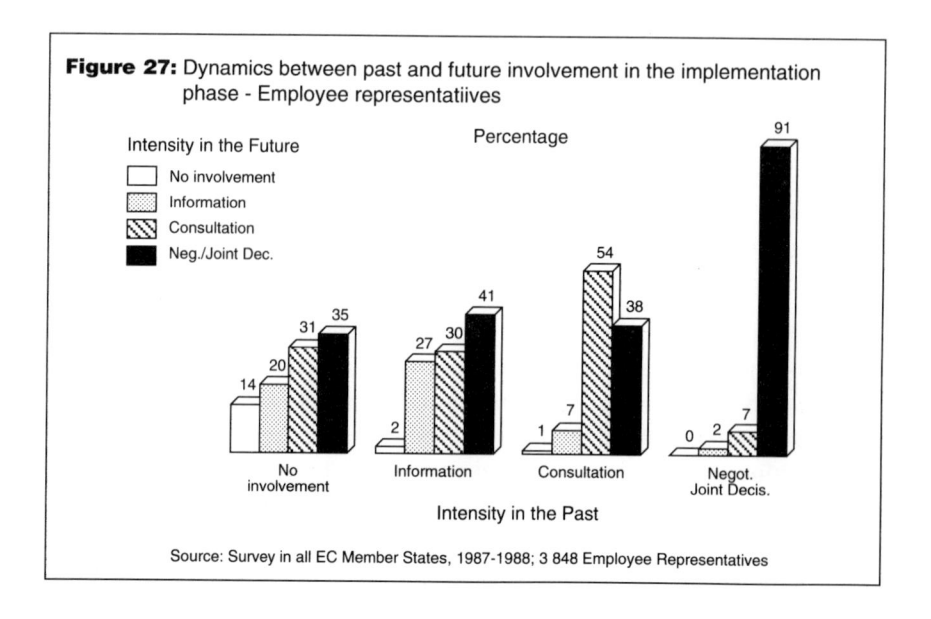

Figure 26: Dynamics between past and future involvement in the planning phase - Employee representatives

Source: Survey in all EC Member States, 1987-1988; 3 848 Employee Representatives

Figure 27: Dynamics between past and future involvement in the implementation phase - Employee representatiives

Source: Survey in all EC Member States, 1987-1988; 3 848 Employee Representatives

percent respectively would be content with such a situation in the future. The vast majority clearly aims to be involved, and their aspirations are far-reaching. For technology planning, a traditional management concern, the intentions stretch evenly over all three intensities of involvement, and about every third employee representative either wants to be informed or consulted or to negotiate or co-determine the situation. Simply to receive information is a rather unpopular concept as well. Only three out of ten representatives want to preserve such a pattern in the future; the trend of aspirations clearly points towards involvement procedures with binding results.

This trend towards consulation, negotiation and joint decision making in the planning phase is virtually repeated in the issue of technology implementation. Yet, in one respect both patterns differ distinctly: In this operational phase in which primary workforce concerns are affected a no-involvement-situation is even less tolerable for employee representatives. Even information provision does not satisfy the needs of representatives who have been excluded from participation in the past. In their own traditional domain, they clearly prefer some form of higher order participation, with 31 percent opting for mainly consultation, and 35 percent indicating a preference for bargaining and co-determination.

When we compare the attitudes of both European managers and employee representatives as to the extent of future participation in the process of technical modernisations of firms, data analysis has shown **similarities of intentions as well as cleavages**. On an aggregate level, both sides are ready to seriously reduce situations of no involvement, to develop participation further and to intensify procedures. But **managers are more hesitant than employee representatives**. On the individual level of comparison which lays bare the dynamics of attitude changes, this result is largely reproduced. This methodological approach shows quite clearly that attitudinal changes of managers are mostly positive for the issue of workforce involvement, but that intended changes towards more elaborated forms of participation are marginal and aim at the next higher form of participation, in principle. At this point, the **largest gap** between the intentions of both sides becomes visible: **Workforce representatives' aspirations are more far-reaching and clearly aim at participation procedures in the way of consultation, negotiation and joint decision making.** This applies to the implementation phase of technology introduction in particular, but these intentions are very evident for technology planning as well. This latter gap might indicate an area of potential discussion and conflict in European companies that introduce new information technology. But given the overall responsive attitude of European managers towards matters of future workforce involvement in technical change, the cleavage does not appear to be large enough to lead to disruptive industrial relations in Europe over such issues.

4. Future participation in management and worforce concerns

So far, the analysis of the future of participation has concentrated on the strategic planning phase and the operational phase of technology implementation. But as to past involvement, we had also investigated further aspects of workforce involvement which centred around content matters. They were called management concerns to denote topics of company decisions which are managements prerogatives by tradition, and workforce concerns which indicated traditional priorities of interest to workers and employees. Data analysis has shown that participation in these substantive matters followed, in principle, the traditional demarcation lines of influence: there was rather little involvement of workforce representatives in decisions which belong to the core functions of managers (like setting market strategies and investment criteria) and large shares of co-determination in matters that center around working conditions. What are the intentions of both sides as to future involvement in these substantive matters?

Figures 28 and 29 serve mainly as a first overview over the intentions both managers and employee representatives have as to the future. The overall message is that among **managers, information and consultation procedures are envisaged in matters of their primary concerns most of all. In matters of workforce primary interests the large majority is ready to involve employee representatives, not only in the way of information and consultation but in co-determination precedures as well**. The split between these intentions and the aspirations of their counterparts is very distinct. A large part of workforce representatives wants to be in involved in management concerns, many of them by co-determination. As to their own substantive interests, the expectations of **European representatives** in regard to being involved are unequivocal: **co-determination is the prevalent attitude when it comes to topics that centre around the working conditions of employees.**

In regard to the results of figures 28 and 29, we want to be very brief in our comments because they do not convey the most interesting information: **how far did attitudes change comparing past experience and future expectations?** Such information can be illustrated at a detailed level by comparing the above figures with figures 3 and 4. Unfortunately, such a procedure creates an information overflow. We have therefore adopted another strategy: we have reduced the single dimensions of both management and workforce concerns to complex measures in order to gain handy tools for both past participation and for future expectations[4]. These complex measures do not render precise information as direct comparisons between the single issues, but they are a useful means to clarify the overall trend in assessments and expectations.

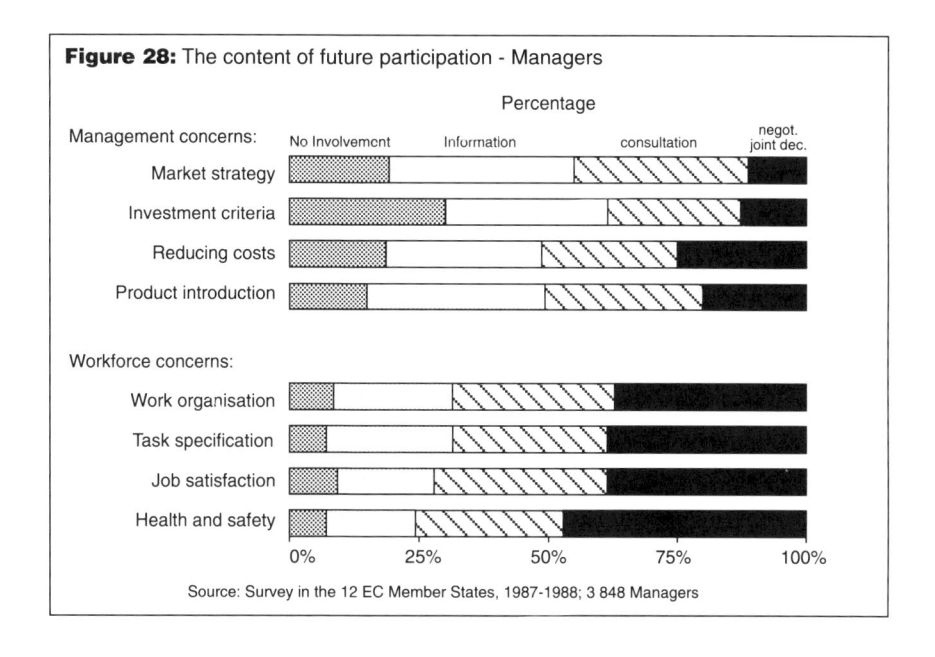

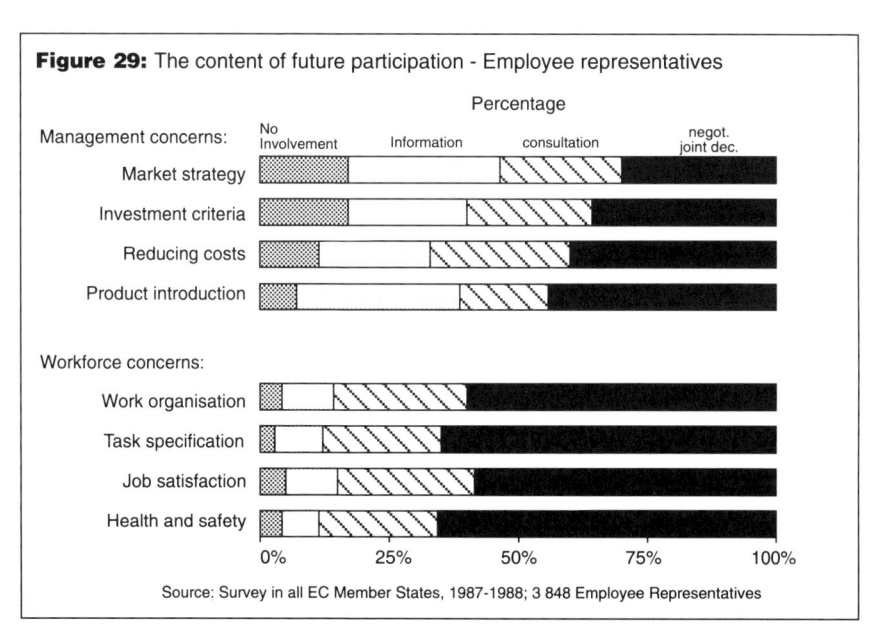

Figure 28: The content of future participation - Managers
Source: Survey in the 12 EC Member States, 1987-1988; 3 848 Managers

Figure 29: The content of future participation - Employee representatives
Source: Survey in all EC Member States, 1987-1988; 3 848 Employee Representatives

Figure 30: Past and future participation in management and workforce concerns - Managers

Source: Survey in all EC Member States, 1987-1988; 3 848 Managers.

Figure 31: Past and future participation in management and workforce concerns - Employee representatives

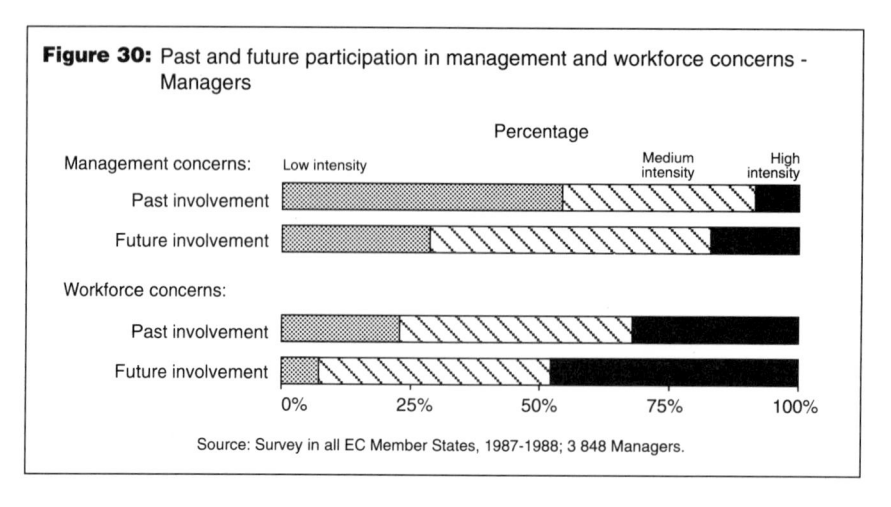

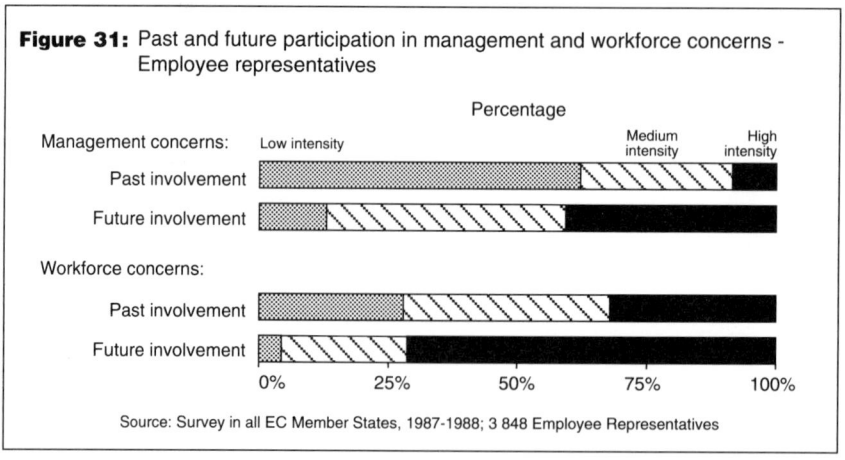

Source: Survey in all EC Member States, 1987-1988; 3 848 Employee Representatives

Figures 30 and 31 tell us that as to participation in content matters, both sides have changed their ideas and expectations considerably. But whereas this **change** can be called distinct in case of managers, it **must be named dramatic when we look at workforce representatives**. Looking at managers first, 50 percent are ready to reduce their past practice of low intensity of workforce representative involvement and move on to participation practices of medium intensity. Even high intensity procedures have doubled. When it comes to workforce concerns which centre around working conditions, managers had exercised rather high level participation to begin with, and they are ready to increase these procedures in the future. This change is expressed by a very small percentage of low future intensity of participation while the share of managers ready for high intensity involvement approaches almost 50 percent.

Employee representatives have quite different expectations as to their future involvement, in management as well as in workforce concerns (figure 31): their involvement in basic company issues, in management concerns, has been very limited in the past. As to this fact, they are in accord with their counterparts on the management side. Whereas management is ready to be somewhat more responsive towards the idea of involvement in their own traditional concerns, workforce representatives have distinct and far reaching expectations: the vast majority wants to be involved in these issues as well. However, they want to be rather strongly involved. Almost two out of five representatives aim at high itensity participation procedures of the co-determination type - twice as much as management is ready to grant. An even larger number wants at least to participate on a medium level of intensity which, in practice, amounts to being informed and consulted. Not to be involved in basic company decisions or at a very poor level only, is tolerated by just a minority of 13 percent. When it comes to working conditions and immediate workforce concerns, employee representatives have very distinct ideas about their future role: 69 percent of the representatives envisage high intensity involvement procedures of the co-determination type. Only every fourth representative aims at weaker forms of participation, and not to participate in matters of direct workforce concerns is envisaged by only 4 percent - a difference which can be explained by measurement errors.

The comparison of past involvement practice and future expectations of participation in substantive company matters reveals, in principle, the **same split between management and workforce representatives which has appeared in regard to participation in the phases of technology introduction:** management has a positive attitude towards stronger involvement of workforce representatives in the future, even in matters which belong to their traditional prerogatives. As to this point, both sides are in accord. The split in attitudes concerns the scope of intensified participation, and here, **workforce representatives** have quite different perceptions as to future necessities: they **envisage much more influence for themselves than management is obviously prepared to concede**.

Further, the analysis of substantive issues of participation reveals a pattern of workforce representative preferences which is closely related to that of future participation in the phases of technology introduction in regard to union strategies at the European level. This strategy aims at high intensity involvement at all levels of company decisions. In issues of direct workforce concerns the wish for participation of the co-determination type can be taken for granted. Such aspirations have a long tradition and have been standard practice in many European companies already. The core of ETUC's latest initiative points towards high level, co-determination type involvement in what was called management concerns: the strategic matters of company decision-making. As was the case with participation preferences in the phase of technology planning, **the majority of European workforce representatives are not ready to envisage such far-reaching participation rights in line with the standpoint of the European Trade Union Confederation.**

5. Conclusions

Reviewing the attitudes and expectations of both sides in regard to future participation, the following results stand out:

- In the European companies, there is an overall positive climate towards increasing participation practices in the future. As a trend, both managers and employee representatives envisage intensities of involvement which are clearly beyond what was practiced in the past. This pertains to participation in the phases of technology introduction and in substantive matters of management and workforce concerns.

- But whereas management plans for rather moderate increases in participation with consultation procedures most of all, workforce representatives have much higher ambitions. They express a clear preference for consultation practices in case of traditional management prerogatives (technology planning and other substantive management concerns). In regard to issues which are close to traditional workforce interests like technology implementation and working conditions, they are stongly in favour of co-determination practices.

These overall results indicate consent and disagreement between management and employee representatives at the same time: consent over the issue that participation practices should be increased and not decreased, dissent on the intensity of future involvement. But given the overall positive climate in the European companies under study which became apparent when discussing the effects of participation (chapter V), the interest cleavages between the two parties do not appear strong enough to prove disruptive in day-to-day company politics.

In the wider context of political strategies and common opinions of both management and union organisations at the European level the survey results indicate that the practice of workforce representative involvement is positively developing towards agreed practices. In Val Duchesse, both employers and unions reached a common understanding that information and consultation should become standard practice in European firms. Such standard practice is clearly underway: not to involve workforce representatives in the future has become a rather rare attitude of European managers. This pertains to participation in the planning and implementation stage of technology introduction as well as in substantive issues of workforce concerns. Only in matters relating to their traditional interests are managers hesitant. Management's attitudes are not only restricted to the practices of information and consultation. In many instances, they are ready for co-determination practices as well.

But compared to the strategic goals of the European Trade Union Conferation which envisages an early, design oriented system of participation of the co-determination type, the cleavage between management's intentions and ETUC's goals is considerable. It is difficult to bridge. Even large parts of union and workforce representatives in the companies, the rank and file of the unions, do not go as far as their own European organisation. Many of them favour co-determination practices, but predominantly in more practical, workforce related matters. When it comes to management interests, traditional demarcation lines are often respected. Most workforce representatives want to be informed and consulted in these issues; co-determination never finds a majority among them. Thus, large parts of the European representatives are in a difficult situation: they want more rights to influence company matters than management is prepared to concede. But vis-à-vis their own organisation at the European level, the majority opt for participation choices which are not in accord with the policies of the ETUC.

NOTES

1. We are aware that this problem is not really relevant to all employee representatives, as not all of them are union members. In our survey, this question had a poor response rate and was answered by 25 % of the workforce representatives only: 49 % were union members, the other 51 % were non-members. At the appropriate points in data analysis, we will point out to a possible difference in reactions of both groups of employee representatives.

2. There is a slight trend for union members to envisage co-determination procedures in technology planning more often than non-members (39 versus 32 %) at the expense of information and consultation procedures which are slightly more often wanted by non-members. At the same time, 15 % of union members do not want any future involvement (only 10 percent non-members). As to technology implementation, non members are slightly more favour of consultation procedures (7 percentage points) while more union members (41 %) than non-members (35 %) envisage co-determination practices.

3. In methodology, this problem is known as the "ecological trap" or as "group fallacy". A prominent example from German political history might serve as a case in point: In the early thirties of this century, the votes for the Nazi party were particularly high in areas with high unemployment rates. The obvious and popular interpretation of this fact that unemployed people favoured the Nazi party was proven wrong by recent social research: It was mainly employed people in regions of high unemployment who were afraid of becoming unemployed themselves and therefore voted for the Nazis.

4. The strategy of data reduction is exemplified for the complex of management concerns: The separate intensities of the four substantive issues (from "market strategy" to "product introduction") received the following weights: (1) for "no involvement", (2) for "information/consultation" and (3) for "negotiation/joint decisions". These weights were added for all four substantive issues which generated a raw index ranging from 4 to 12. "4" is the lowest number of counts and means "no involvement in all four issues" while "12" signalises the fact that respondents voted for co-determination procederes in all four issues. By combining the values 4 - 6, 7 - 9 and 10 - 12, we arrived at the three categories of low, medium and high participation intensity.

Chapter Seven

Country Specific Analysis

1. Introduction

In this section of the Report we examine the subject of employee participation in technological change in the European Community in each of the Member States. We will first give an overview of the extent of past involvement in each country and seek to explain the wide variations in the various forms of employee participation in the introduction of new technology that are evident from the survey data. Secondly, the data reveals distinct differences from one country to another on the preferences of both employers and employees for the nature and extent of employee participation that they have for the future. Finally, we compare the data on past participation with the expectations of both parties for the future and comment on the implications of the survey results in the light of the Val Duchesse Social Dialogue of March 1987. The results are evaluated with respect to the policies on employee participation in technological change both of UNICE, The European Employers' Federation, and the ETUC, the European Trade Union Confederation.

In order to explain the data in this survey, it is important to identify the **factors** which assist or impede the possibilities of employee involvement in the introduction of new technology. By identifying such factors, the individual country differences within the member states of the European Community which appear in the data in this survey can be much more readily accounted for, although the survey data does not always contain enough detailed information to relate the country differences to every one of these factors in every case.

2. Country specific factors affecting involvement in the introduction of new technology

There are a number of country-specific factors which shape the opportunities for employee involvement in the introduction of new technology.[1] It is important to mention at the start that some of these factors will not only be country-specific but will also apply at sectoral level and company level in the countries themselves. The factors will also be inter-related to some degree, and strong forms of particular factors might well have a dominating impact on the outcome of the degree of employee participation in technological change that emerges. These factors will not only shape the opportunities for participation but they will also affect the subject matter and timing of involvement. There are five principal factors which can help or hinder the opportunities employee representatives have for participation; they will not, however, determine the extent to which employee representatives can influence the effects of technological changes in the organisation. It is one thing to put discussion about technological change on to the formal agenda and quite another to translate these participative opportunities into a real ability to influence the way new technology can affect working life.

The extensive surveys and case studies that have been carried out in several countries in the European Community during the 1980s suggest that the opportunities for employee involvement in decision-making about new technology depend on the following five main variables:

(a) management's dependence on the skills and co-operation of its workforce to achieve its objectives for introducing the new technology;

(b) management style and its attitude to participation;

(c) the bargaining power of organised labour to force management to negotiate or consult with its representatives in the absence of any voluntary disposition on the part of management to do so;

(d) regulations which lay down participation rights for employees or their representatives on a range of matters at enterprise level;

(e) the degree of centralisation of the industrial relations system that exists in the country concerned;

(a) Management's reliance on its workforce to achieve its objectives

As we saw in Chapter 2 of this report, the objectives that management has for introducing new technology will obviously vary from one organisation to another. In most cases management will have several goals in mind when introducing new technology into their organisation. The emphasis between these is likely to vary according to the priorities and purposes of their organisation and the context in which it operates. These goals include increased competitiveness, greater control over the work process, reduced costs, enhancing product quality, improving customer service etc. Whatever management's goals, the nature of technological innovation is of particular importance.

For example, companies which concentrate their activities either on the sale or production of standardised goods or services, which operate within mature price competitive markets or where office technology is used to reduce costs by standardising the collection or processing of information will utilise process innovations. The effect of such technological change in cases such as this is likely to lead to a more rigid segmentation of work and a greater concentration of programme-related functions which are carried out by specialist departments. In such circumstances, there is little need to involve their employees or their representatives in the planning or implementation of technological change unless they are forced to do so by the threat of industrial disruption or a reluctance by the workforce to accept the changes.

Conversely, where management is concerned to enhance the competitiveness of their organisation by offering its customers improved levels of service and enhanced product quality, the new technology introduced will serve to highlight product performance, quality control, design requirements, technical sophistication, more flexible scheduling and greater emphasis on customer requirements. Such flexible systems require more adaptable and broadly skilled employees who are adept at problem-solving and whose skills and co-operation are highly valued.

Thus, management's dependence on its labour force for realising its objectives is an important factor in influencing the propensity shown by management to pursue a participative approach to change. The less management relies on the skill and expertise of computer specialists and senior managers and the more it relies on the skills, co-operation and problem-solving capacity of lower-level personnel, the greater the prospects of participation in technological change. For example, the premium which has been increasingly placed on maximising the commitment of employees to the "culture" of the company for which they work in the **United Kingdom** throughout the 1980s by the adoption of methods such as team working, briefing groups, quality circles, employee reports, company presentations within devolved organic structures coupled with the growth in joint consultation committees has led to considerably more participation of an informal kind in British companies.

It is important to qualify all this by making a number of observations about the form of participation, its content and its timing. Participation will usually be of an **informal** kind, largely involving discussions with individual workers about job-related issues. Management itself will determine the stage at which employees are involved - usually at the early part of the **implementation** stage. Moreover, participation will involve **consultation** rather than negotiation. The parties, agenda and timing of participation is thus of management's own choosing.

(b) Management style and attitude towards participation

The pre-existing style of management will also have an important bearing on whether or not technological change is introduced by participative means or by managerial fiat. Traditions of co-operative industrial relations are marked by consultation and participation; a history of mistrust and antagonism is associated with an unwillingness on the part of management to have its prerogative challenged when it introduces technological change.

Whilst management attitudes, values and ideology can sometimes change in response to outside influences such as competitive product market pressures and the nature of the technology being employed, such an adaptation of managerial practices to environmental pressures can be difficult. In companies where managerial aversion to employee participation exists it tends to permeate through the whole management structure. At each level in the management hierarchy, the approach that individual managers take to employee participation will reflect the way they themselves are controlled from above. In such circumstances, any participation that does occur will be complementary to management's right to manage and make decisions, and not a challenge to that power.

Unless there is a very strong tradition of co-operation with employees' organisations, participation will be limited to employees rather than employee representatives and to involvement in task-related matters rather than strategic matters. Indeed, management may offer involvement by employees in job-related issues specifically to avoid the involvement by trade union or employee representatives. If employee representatives are admitted to the decision-making process it is likely to be at the **implementation** stage when the major design questions have been decided. Moreover, it is unlikely that the participation of employee representatives will be extended to issues such as the pace of work, job design, manning etc. let alone major strategic issues such as technological choice and investment.

There are wide variations in management style throughout the European Community. These range from the co-operative management styles which are generally found in Northern European countries, particularly Scandinavian countries such as **Denmark**, and also to a lesser extent in **Germany,** the **Netherlands** and **Belgium** to those in Southern European countries such as **France, Spain, Portugal** and **Greece.** Clearly, this is no more than a generalisation, and there are, of course, great variations from company to company, region to region, and sector to sector in each country.

(c) The bargaining power of organised labour

An obvious factor which determines the extent to which trade unions and employees' organisations can gain access to decisions over technological change and new systems of work organisation is their actual power to force management to consult and bargain with them. This power will in turn depend on a number of other factors, the most important of which include membership density, the willingness on the part of employees to act collectively in defence of their interests and their capacity to inflict damage on their employer. In some countries, of course, some of the trade unions themselves as a matter of policy deliberately refrain from being involved with management in technological change and sometimes this means they refuse to bargain about such change as well. Examples of such hostile attitudes to participation can be found among some unions in **France** and **Italy**. Survey evidence from the **United Kingdom** shows that trade unions are more likely to use what bargaining power they have when technological change is accompanied by major organisational change.

It is important to understand that the issues which are thrown up in the workplace as a result of technological change do not readily lend themselves to the formal agenda of collective bargaining. The complexity and differential impact of changes in technology will often make it difficult to forecast its precise effects on skills, job content and job security. Moreover, its impact will vary from one job to another. Some employees may see the danger of being deskilled or perceive that their job security is being threatened whilst others may see themselves benefiting in terms of wider job responsibility, better promotion prospects or improvements in pay. In circumstances, therefore, where a trade union is facing technological change where its members are affected differently, it is difficult to present a united approach to management in the bargaining process.

The evidence from a number of studies[2] suggests that the degree of overall influence exerted by organised labour will be affected by five inter-related factors; these include the depth and strength of a union's organisational arrangements within the enterprise, the technical knowledge and expertise of its membership, the resources that it is able to apply to the task of developing alternative technological options, the ability to develop detailed strategies to guide and assist its membership in the process of change and finally, the capacity to threaten or possibly use sanctions to induce management to compromise on its proposals.

British experience shows that the existence of multi-unionism at enterprise level can considerably limit a united response to management. Because each group is represented by a different trade union, inter-union relations can be soured as a result and it is difficult for unions to bury their inter-union and inter-occupational differences. Moreover, it may also enable employers to mobilise the support of certain occupational groups and their unions who see opportunities to profit from the adverse effects of new technology on others. Hardly any of the New Technology Agreements which were negotiated in the **United Kingdom** during the early 1980s had more than one union signature on them. In countries where labour movements have been largely successful in implementing the principle of 'one plant one union' - for example in **Germany** - conflicts about contradictory interests between occupational groups are not absent but they can be dealt with within a single representative structure.

Scandinavian experience illustrates that a key factor in the ability of unions to influence the direction of change brought about by new technology has been the educational support programmes for their membership at local level and access to research and development information independent of employers. This has involved a conscious decentralisation policy on the part of national union leaders to involve shop floor members in technological change by drawing on their knowledge and skills and giving them support when they need it.

There are two critical points in the process of technological decision-making where organised labour needs to be represented: at the company level where the key investment decisions are made and questions of equipment selection and design are resolved, and at the establishment or workplace level where those decisions are implemented. An effective organisational structure, therefore, is one which enables both the negotiation of broader strategic issues at the company level and more detailed and specific issues at a localised level. In multi-establishment companies decisions to introduce new technology - particularly when it has a negative effect on earnings or manning levels - is likely to be centralised. Therefore, organised labour needs a structure which can support negotiations at the **planning** stage of the introduction of new technology; such a structure is essential for success in negotiations. However, employees' organisations can also achieve important concessions from management by negotiation during the **implementation** phase of the technology at the workplace. The limited attention paid by central decision planners to the labour aspects of new technology can sometimes provide considerable scope for local variations in matters such as staffing levels, the content and design of jobs, skills training, arrangements covering manning levels and the extent to which companies can enforce an intensification of work effort.

(d) Regulation

An important factor which is likely to shape the opportunities for employees to participate in technological decision-making is where there are rights in existence which oblige employers to provide their employees with appropriate information and to consult and/or negotiate with them. These rights take various forms and vary according to the legal systems of the countries concerned. They may be defined in the constitution, specifically laid down by statute, incorporated in collective agreements which have legal effect or may simply be based on jurisprudence or common law.

Two points need emphasising in relation to the underpinning of participation by regulation. First, regardless of the form that such regulation takes there is **no guarantee** that effective participation will necessarily result from it. Second, the rights of participation that exist may not necessarily be found in any legal code; in some countries - for example the **United Kingdom** - the exercise of information and consultation rights can sometimes be subsumed under conventional collective bargaining without any legislative back-up. Moreover, not only is there a great deal of diversity in the regulatory provisions for participation from one country to another, but also there may be differences between enterprises and industrial sectors within the same country.

In the majority of member states of the European Community a clear distinction is made between rights associated with **collective bargaining** (which may or may not be underpinned by legislation) and rights concerned with information, consultation and participation in enterprise decision-making which are often exercised under the auspices of **separate bodies** which deal with matters other than the negotiation of terms and conditions of employment. These separate bodies usually exist at the level of the company or the establishment and they often provide for participation in technological decision-making. Examples of such separate bodies are Betriebsrat (**Germany**), Comité mixte d'entreprise (**Luxembourg**) and Semarbijdsudvalg (**Denmark**). It is important to stress that the opportunities for employees to be involved in participation in decisions about new technology are shaped by **both** collective bargaining and any separate participatory body that might exist. Indeed, there is a close relationship between the two.

(e) The degree of centralisation of the industrial relations system

Regulation on participation in new technology is also closely related to the political traditions in the country concerned which have shaped the industrial relations system. Countries where negotiations are centralised above the shopfloor level, i.e. are conducted for entire companies, industries, or cover the whole labour market tend to have a greater degree of industrial democracy. For example, **Scandinavian countries** and **Germany** have built on a long history of participatory management schemes. These traditions have developed within a general industrial relations context commonly described as "corporatist".

"Corporatism" involves collaboration between capital and labour in pursuit of joint goals, such as economic growth, which are perceived as the common basis for successful pursuit of their common interests, where both sides may moderate their demands so as not to jeopardise their joint goals. The State will seek to actively co-ordinate this process by the devolution of some State authority to organised interest groups, especially for the purpose of governing their constituents. In order for this State devolution to be effective, the interest groups must be organised comprehensively, with an internal concentration of private government, and representational monopoly granted by the State as a quid pro quo for the moderation of demands. It is important not to over-stress the formal nature of corporatism; in practice some of the strongest forms of corporatism occur in an emergent and informal way and do not necessarily require deliberate action on the part of any of the participating parties to create formal structures in which corporatism takes place. Corporatism works best when both employers and trade unions are organised centrally i.e. when they are able to exercise authority over their respective affiliated organisations.

Within the European Community, both **Denmark**, **Germany** and the **Netherlands** - and to a much lesser extent **Belgium** - fulfil these criteria and such traditions of "corporatism" have clearly contributed significantly to the planned introduction of new technology in these countries. Corporatist industrial relations practice have provided a structure and framework, as well as a stock of necessary goodwill and centralised bargaining experience on all sides.

Generally speaking, we would expect to observe **higher levels of employee representative involvement, less resistance to, and more consensus on technological change** than would be found in countries which had more decentralised industrial relations systems. The **more** the industrial relations system is organised on **strong central associations** which enables the two sides to bargain at all levels within a central framework, the **more effective** the rules of participation will be.

In the preceding section of this report we have identified **five** variables which would appear to play an important part in shaping the opportunities for employee involvement in the process of technological decision-making: **management's reliance on its workforce** to achieve its objectives for introducing the new technology; **management style** and its attitude to participation; the **bargaining power** of organised labour to force management to negotiate or consult with its representatives in the absence of any voluntary disposition on the part of management to do so; regulatory provisions which lay down **participation rights** for employees or their representatives on a range of matters at enterprise level; and finally the degree of **centralisation of the industrial relations system** in the particular country (see **TABLE 1**). Whilst not all these criteria can be used as a basis against which to assess the survey data on individual countries, they are clearly important in the overall debate on participation in technological decision-making.

TABLE 7.1
Factors affecting employee participation in technological change

Variable	*Favourable conditions*	*Unfavourable conditions*
Technological Objectives	Performance enhancement and problem-solving skills important for success	Cost reduction with little dependence on employees
Management Style	Co-operative	Conflictual and closed
Bargaining Power	Highly unionised; facing common technological threat, and strategically located; technically knowledgable and skilled membership; united and cohesive union organisation	Multi-unionism, low unionisation, facing uncertain or variable impact from technology; lack of research resources; inexperienced officials; unions divided along political or religious lines.
Regulation	Established traditions of tri-partite corporatism strong forms of law	'voluntaristic' 'market forces' or weak forms of legislation
Industrial Relations System	Centralised collective bargaining agreements	Decentralised collective bargaining agreements

NOTES

1. Much of the discussion that follows is adapted from Stephen Deery, 'Determinants of trade union influence over technological change', New Technology, Work and Employment, 4, 2, 1989.
2. See Batstone E. et al. Unions, Unemployment and Innovation, Blackwell, 1986; Clark, J. et al. The Process of Technological Change, CUP, 1988 and Gustavsen, B. 'Technology and Collective Agreements: some recent Scandinavian developments', Industrial relations Journal, 16, 4, 1985.

CHAPTER EIGHT

A Comparison of Past and Future Participation in Technological Change in the Twelve Member States

The factors which shape the involvement of employees and their representatives in decision-making about the introduction of new technology which were outlined in the previous section lead us to expect that there will be a greater degree of participation in both the planning and implementation stages of technological change in some EC countries rather than others. We would be most surprised, for example, if **Denmark** and **Germany** did not register the highest degree of participation of all the twelve member states in both the planning and implementation stages of technological change.

Denmark has built on the experience of "data" or "technology" agreements which originated in Norway in the late 1960s and 1970s and the unions there sought a more direct influence upon technological development at work, by negotiating such agreements centrally with Danish employers. Unlike Sweden, which embarked on legislation to regulate technological change, the Danish employers' organisation (DA) was prepared to negotiate agreements which imposed a duty on employers to inform the employees about any major technological change "in due time" before anything was carried out. **Denmark** also has employers who pursue a co-operative managerial style and who view the trade unions as social partners. In addition, the Danish industrial relations system is highly centralised and "corporatist" in character, there is a high dependence by employers on the co-operation and skills of organised labour, and Danish trade unions have a great deal of bargaining power.

Germany has a semi-centralised sectoral bargaining system, extensive legislation on "co-determination", a high dependence by managers on the skills and co-operation of the workforce, a fairly co-operative managerial style; however, although since 1972 the works councils have been able to have union representatives present at their discussions, German union representatives cannot bargain at workplace level except via the works council, which is in contrast to Denmark.

In both **Denmark** and **Germany**, therefore, there are favourable factors in existence which assist employee participation; in both countries there are long traditions of co-operative management practices with trade unions underpinned by a system of tri-partite consultation and state support for jointly agreed goals between the two parties. Trade unions are accepted by management as partners in the planning of technological change and there is a stock of necessary goodwill between both sides of industry, although in some German hi-tech sectors labour is non-union. Management in both countries uses new technology to increase their competitiveness, to enhance product quality and improve customer service; in such cases we would expect a high degree of dependence by management on the co-operation of the workforce in problem solving etc. In addition, there are extensive provisions in both countries for training and retraining in the use of new technology coupled with effective means for re-deployment, time off for employees to perform their duties and with a premium placed on job satisfaction and adequate job design.

Similarly, we would expect that EC countries with a history of trade union repression by the state in the post-war period and where trade union organisation is fragmented and badly developed to have the lowest degree of participation in technological change. In these countries management has been used to exercising its prerogative with little, if any, consultation with employees e.g. **Portugal**, **Spain**, **Greece**. The trade unions in each of these countries only received legitimacy in the mid-1970s, collective bargaining is poorly developed and trade union organisation in the cases of **Portugal** and **Spain** is split along political and/or religious lines. In **Greece** there was a recent split in the main trade union confederation, the GSEE.

There are also countries in the European Community where management is determined to maintain its prerogative and is reluctant to embark on higher levels of participation. The trade unions in these countries in turn are reluctant to become involved in any participation schemes which compromise their traditional reactive, oppositional role in the workplace. In such countries the union movement is fragmented along political and/or religious lines and there is a great deal of rivalry between a number of competing trade union confederations. In such countries the industrial relations system tends to be decentralised and conflictual. On the other hand, there may be an increasing dependence by management on the skills and co-operation of the workforce as the new technology takes shape; there may also be a degree of favourable legislation as the state seeks to play a conciliatory role between employers and trade unions and unions have a fair degree of bargaining power either in the industrial or political arena. We would expect countries generally falling under this classification (e.g. **France** and **Italy**) to appear in the middle band of our participation table. However, although trade unions in both these countries eschew participation, the nature of bargaining power at company level differs. Since the mid-1970s Italian bargaining (at least in much of the country) has become well established, while in France unions have remained weak in bargaining and increasingly dependent on "political" protection from the State.

Given our discussion in the introductory section on the factors which determine the degree of participation in the process of technological change, we would also expect that there would be a much greater degree of participation in the **implementation** rather than the **planning** phase of technological change in all countries. As we saw in our introductory discussion to this section, in many cases where technology is introduced into enterprises management will depend on employee co-operation and goodwill in the implementation phase in order to effect a successful change-over to the new system of working. However, there is little need to involve employees in the planning phase unless management is required to do so either by legislation or as a result of a collective agreement with the trade union concerned; in such cases the strategic objectives of management in introducing new technology and the choice of technology to meet these objectives is solely the prerogative of managers and engineers.

Are such expectations about the different degree of participation by employees in technological change in the 12 EC countries reflected in the survey data? Figures 32 and 33 give a breakdown by country of the extent of past participation in the **planning** phase of technological change as given by management and employee representatives respectively.

Tables 32 and 33 show that **Denmark** and **Germany** have the highest level of involvement by employee representatives in the planning phase of technological change, with both countries having the lowest degree of "no involvement" and the highest degree of "negotiation" or "joint decision-making". As one would expect, **Portugal**, **Greece** and **Spain** also come at the bottom of our table with the highest number of respondents who reported "no involvement" by employee representatives in technological change and the lowest levels of negotiation or joint decision-making. Ireland, the **Netherlands** and **Belgium** also rank in the upper part of our two tables. On the one hand, the industrial relations system in **Ireland** has become increasingly centralised in the last two decades, and the trade unions there have a fairly high degree of bargaining power. On the other hand, there is very little legislation promoting participation[1] in **Ireland**, Irish managers do not favour participation and there is a low degree of management reliance on the skills and co-operation of its workforce. In the **Netherlands** and **Belgium** there is extensive consultation at national level before any employment legislation is enacted and both countries have well-established works council systems with extensive rights given to their members.

A striking feature of Figures 32 and 33 is the high levels of "no involvement" in all 12 EC member states reported by both sides; only in the cases of **Denmark** and **Germany** do we find less than one in four managers and employee representatives who state that there had been no involvement of employee representatives in the planning phase of technological change. It is also evident that there is much disagreement between managers and employee representatives about the different levels

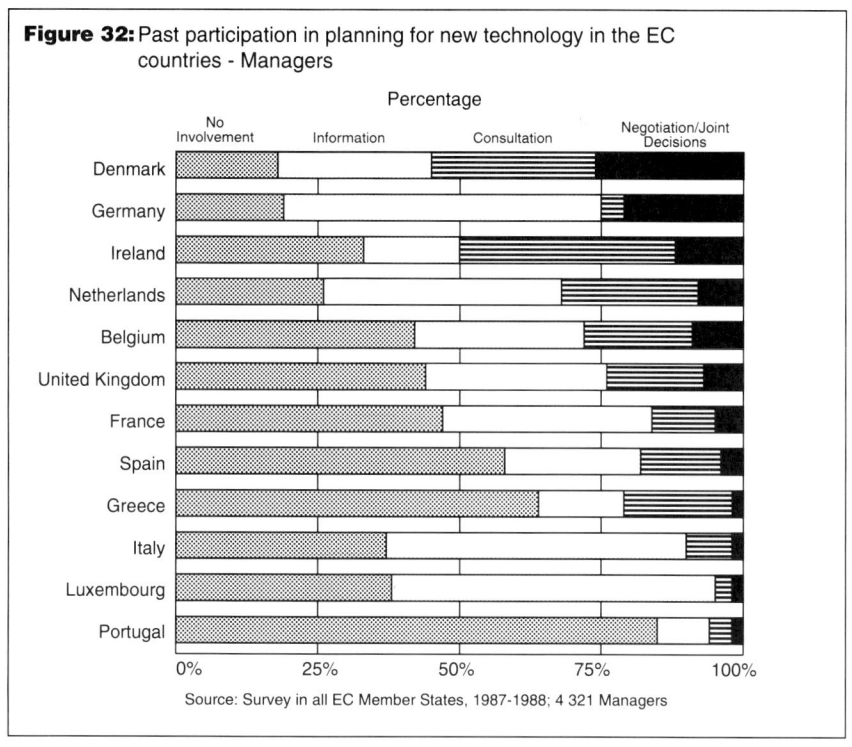

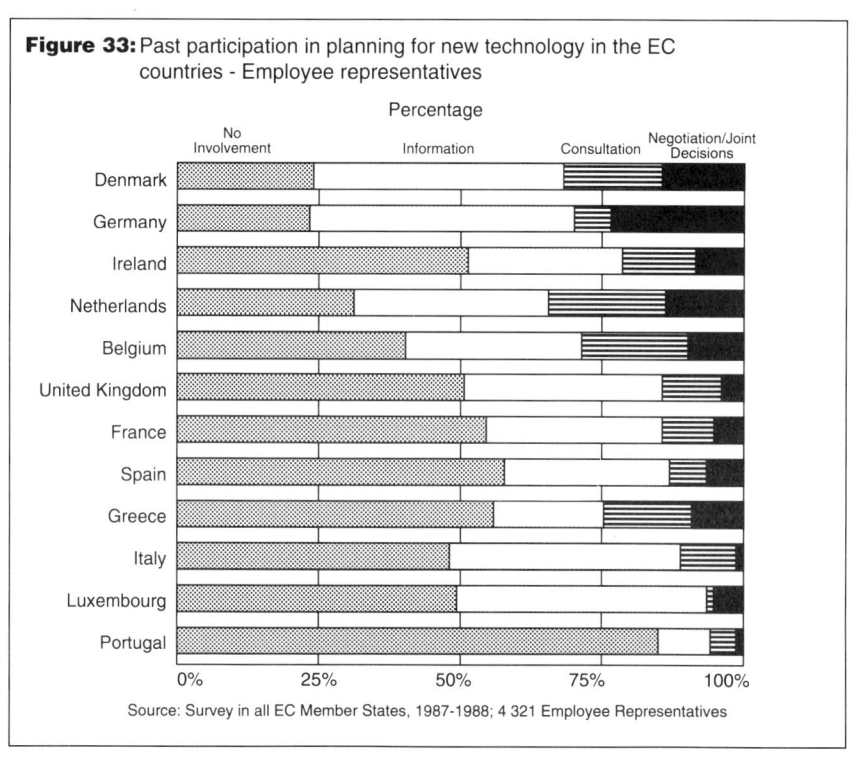

A Comparison of Past and Future Participation in Technological Change

Figure 32: Past participation in planning for new technology in the EC countries - Managers

Source: Survey in all EC Member States, 1987-1988; 4 321 Managers

Figure 33: Past participation in planning for new technology in the EC countries - Employee representatives

Source: Survey in all EC Member States, 1987-1988; 4 321 Employee Representatives

of past participation in two countries: **Denmark** and **Ireland**. Managers in **Denmark** claim that there is a higher degree of consultation and negotiation or joint decision-making than their employee representative counterparts; this is also the case in **Ireland**. Only in **Denmark** do we find a *majority* of managers reporting that consultation, or stronger levels of involvement, occurs. Moreover, this result in *contradicted* by Danish employee representatives (see Figure 33).

Figures 34 and 35 give similar information on past participation of employee representatives in the **implementation** phase of technological change. If we compare the results of Figures 32 and 33 with those of 34 and 35, we can see that in all 12 member states the extent of "no involvement" is lower in the implementation phase than in the planning phase and the degree of "negotiation" or "joint decision-making" is higher in the implementation phase. This is not at all surprising and it is very much in line with what one would expect from the European-level analysis. We also find that the ranking of countries in the planning phase of technological change bears a close similarity to that in the implementation phase, with employee representatives again proving to be more sceptical about the level of employee representative involvement than their management counterparts. The main differences between managers and employee representatives are most prominent in the percentages of the two sides who claimed that there was "no involvement" (**United Kingdom, Greece, Italy** and **Luxembourg**) and levels of participation involving consultation or higher levels (**Denmark, Ireland,** and the **United Kingdom**). **Denmark** and **Germany** again have the highest degree of participation in both phases. It is also noticeable that three countries register substantially higher numbers of both sides who stated that there was consultation in the implementation phase than in the planning phase (**Ireland, United Kingdom** and **Greece**).

The figures also illustrate that our expectations of low levels of participation in technological change in **Portugal**, **Spain** and **Greece** are also borne out by our survey results. Managers in all three countries stated that "negotiation" or "joint decision-making" took place very rarely in the planning phase of technological change (**Portugal** 2%; **Spain** 4% and **Greece** 2%); in the implementation phase the percentages were also low (**Portugal** 5%, **Spain** 9% and **Greece** 16%). There is one country which seems to stand out in our results; **Portugal** has the **highest** level of "no involvement" in the **planning** phase of all European Community countries with both managers and employee representatives in close agreement that this is indeed the case (85% and 84% respectively in Figures 32 and 33). However, the results for **Portugal** are very different in the **implementation** phase (see Figures 34 and 35); here the results

A Comparison of Past and Future Participation in Technological Change

Figure 34: Past participation in implementing new technology in the EC countries - Managers

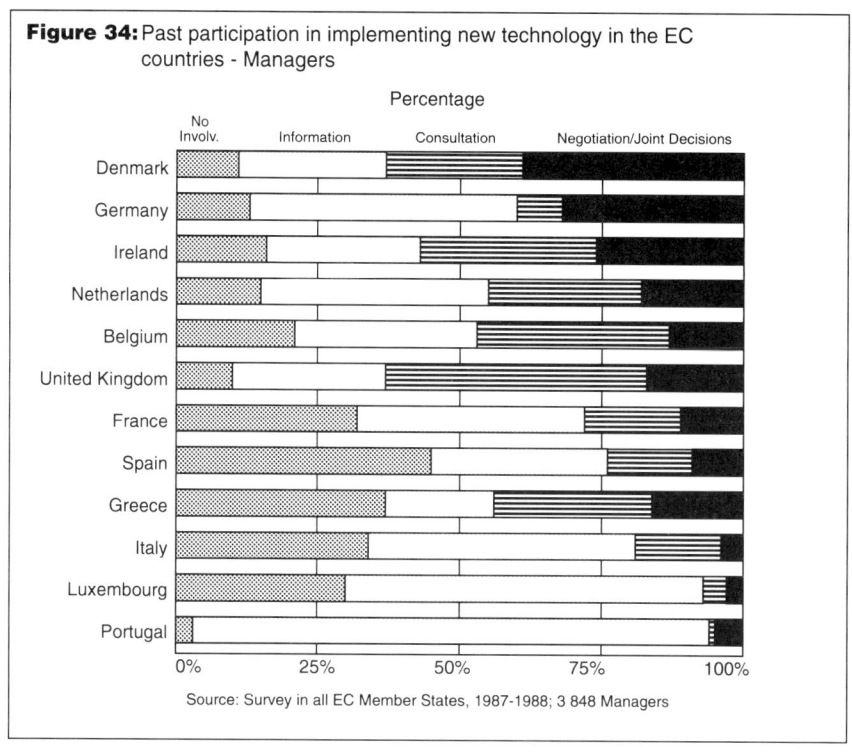

Source: Survey in all EC Member States, 1987-1988; 3 848 Managers

Figure 35: Past participation in technology implementation in the EC countries - Employee representatives

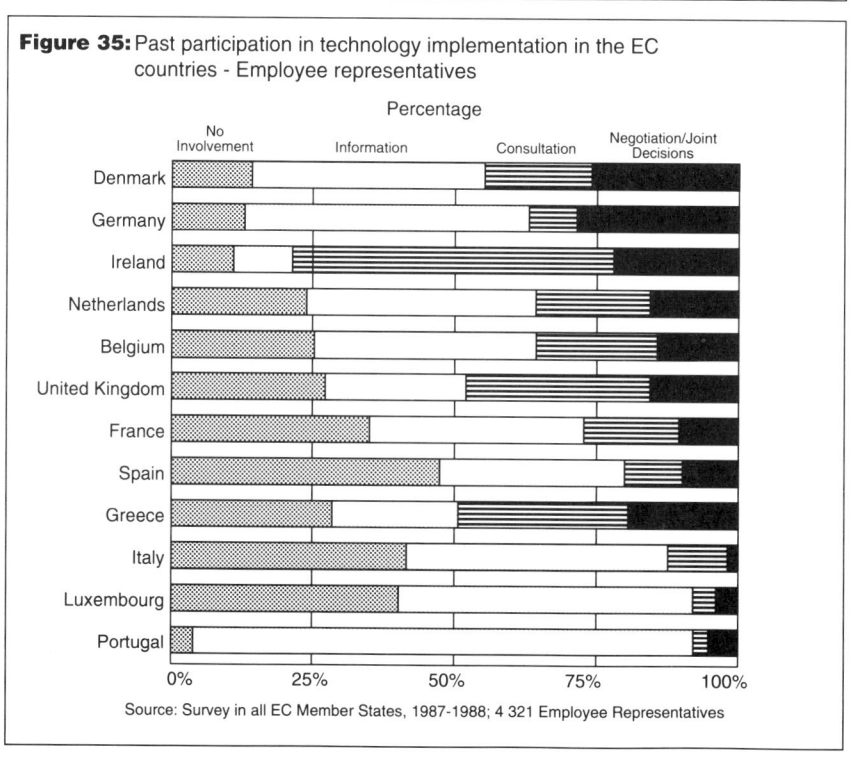

Source: Survey in all EC Member States, 1987-1988; 4 321 Employee Representatives

for "no involvement" are the **lowest** of all the countries in our survey (3% managers and 4% employee representatives) with a very high level of involvement consisting of the provision of information to employee representatives about the implementation of technological change. As we will see later in our separate discussion for this country, this is largely due to the very detailed and precise legislative provisions which define the rights for information and consultation for works councils in Portugal.

Figure 34 shows that only in **Denmark, Ireland** and the **U.K.** do we find a *majority* of managers who state that the **implementation** stage of technological change is characterised by at least consultation or higher levels of involvement. Of these three countries, only in **Ireland** do we find that these managers' opinions are not contradicted by employee representatives (see Figure 35).

What are the **future** intentions of managers and employee representatives in the European Community countries on employee representative involvement in technological change? Figure 36 and 37 show the preferences of the two sides for the kind of participation they would like in the **planning** phase of technological change in the future, given their past experience. The most striking conclusion that can be made is that there is a **significant shift of opinion among both sides in favour of a greater level of participation in all the countries surveyed**. It is also evident that **employee representatives would prefer a higher level of participation than their management counterparts**; this is only to be expected. Thirdly, the **pattern of preferences for the future, in terms of the ranking of countries, bears a close resemblance to that of the past.**

What is particularly evident in comparing Figures 36 and 37 is that there is a high degree of disgreement between the two sides in their preferences for the type of participation that they would like in the future. This is true in all European member states but is particularly striking in **Denmark** and **Germany** — our two most highly-ranked countries. It is clear that employee representatives, having experienced participation in one form or another during the 1980s, are keen to have a substantial role in the strategic decision-making that takes place when new technology is being introduced into companies. This spells out a clear message for the future, and might possibly herald conflicts between the two sides in these countries. Since the beginning of the 1960s both these countries have been noted for their low levels of industrial conflict, and even though the industrial relations systems in both countries differ substantially, both countries have been noted by commentators as having stable, conflict-free and co-operative relations between both sides of industry, although during the 1980s there is evidence that both these countries are more conflict-prone than before.

A Comparison of Past and Future Participation in Technological Change

Figure 36: Future participation in planning for new technology in the EC countries - Managers

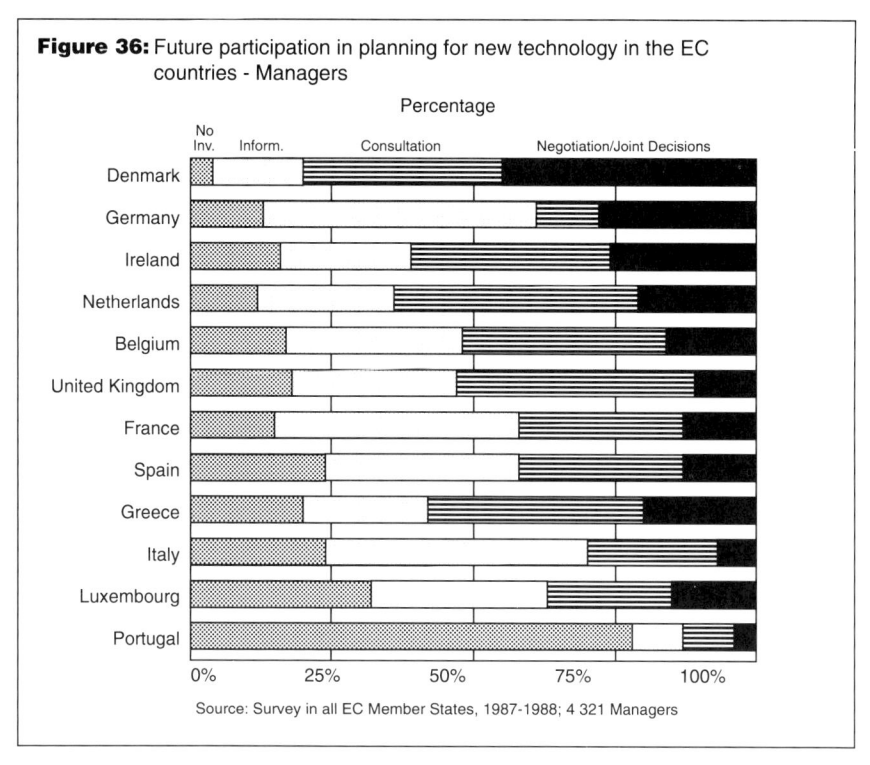

Source: Survey in all EC Member States, 1987-1988; 4 321 Managers

Figure 37: Future participation in planning for new technology in the EC countries - Employee representatives

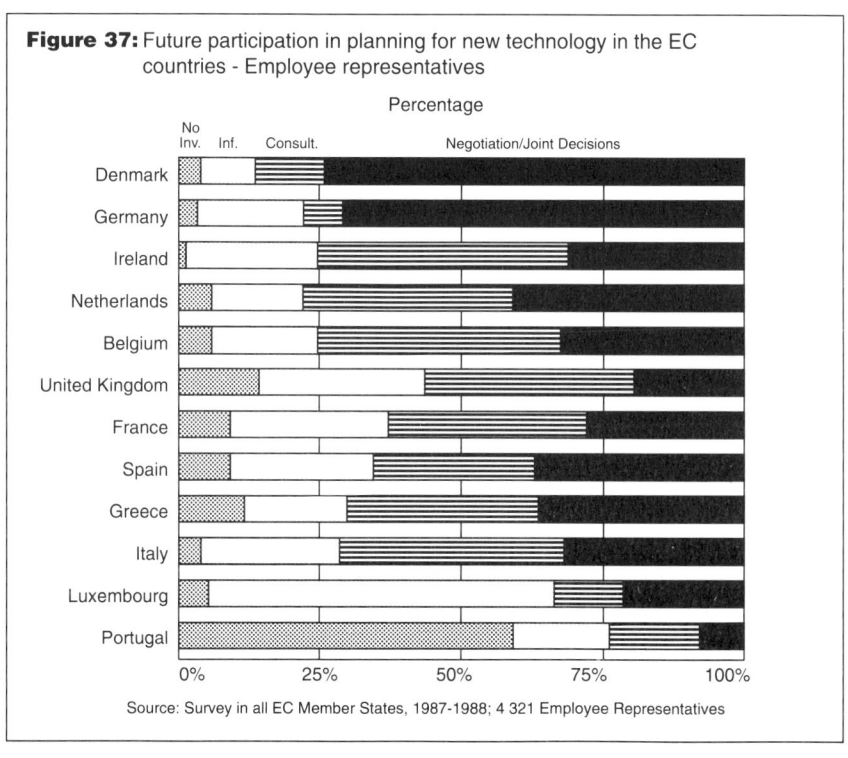

Source: Survey in all EC Member States, 1987-1988; 4 321 Employee Representatives

In the **planning** stage of technological change there is a significant improvement in the preferences of the two sides for future participation. **Denmark, Greece, Ireland,** the **Netherlands,** and the **U.K.** all have results which show that a *majority* of managers intend to have levels of participation involving consultation and above. A *majority* of employee representatives in all countries except **Luxembourg** and **Portugal** would favour levels of participation involving consultation and/or "negotiation" or "joint decision-making".

Figures 38 and 39 show a similar pattern in the **implementation** stage to that of the planning stage insofar as the future preferences of both sides are concerned. Again we can see that a similar ranking of countries is evident, with **Denmark** and **Germany** showing the highest preference for strong forms of participation. There are also differences between managers and employee representatives in their future preferences with a difference between both sides in **Denmark** and **Germany. Another feature of Figure 39 which stands out is that the degree of no involvement shrinks in nearly all countries to below 5%.**
A *majority* of managers in all countries except **Italy**, **Luxembourg** and **Portugal** envisaged levels of participation in the future which consisted of participation levels involving consultation or stronger forms (see Figure 38). A *majority* of employee representatives in all countries except **Luxembourg** and **Portugal** opted for participation levels involving consultation and above.

In the next section of the report we consider the results for individual countries and attempt to interpret them in the light of our explanatory framework. As we have seen, our results show that there are wide variations in participatory practices across the member states of the European Community. In each country we first outline the nature of the industrial relations system and then seek to identify factors which impede or promote participation in technological change in the country concerned. In each case we seek answers to the following questions:

1. To what extent does our survey show that the provisions of the Val Duchesse Joint Opinion between UNICE and the ETUC of March 1987 are being fulfilled in the planning and implementation stage of technological change in each member state?

2. To what extent does our survey show that both sides intend to fulfil the provisions of Val Duchesse in both stages of technological change in the future?

3. To what extent is there agreement between managers and employee representatives on the degree of participation desired in the future?

A Comparison of Past and Future Participation in Technological Change

Figure 38: Future participation in implementing for new technology in the EC countries - Managers

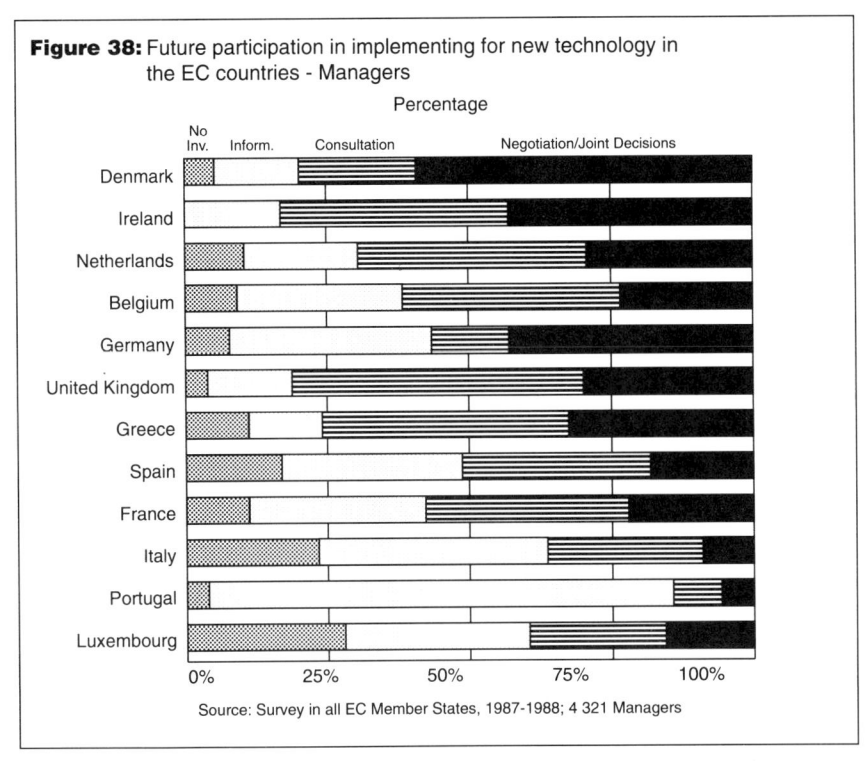

Source: Survey in all EC Member States, 1987-1988; 4 321 Managers

Figure 39: Future participation in planning for new technology in the EC countries - Employee representatives

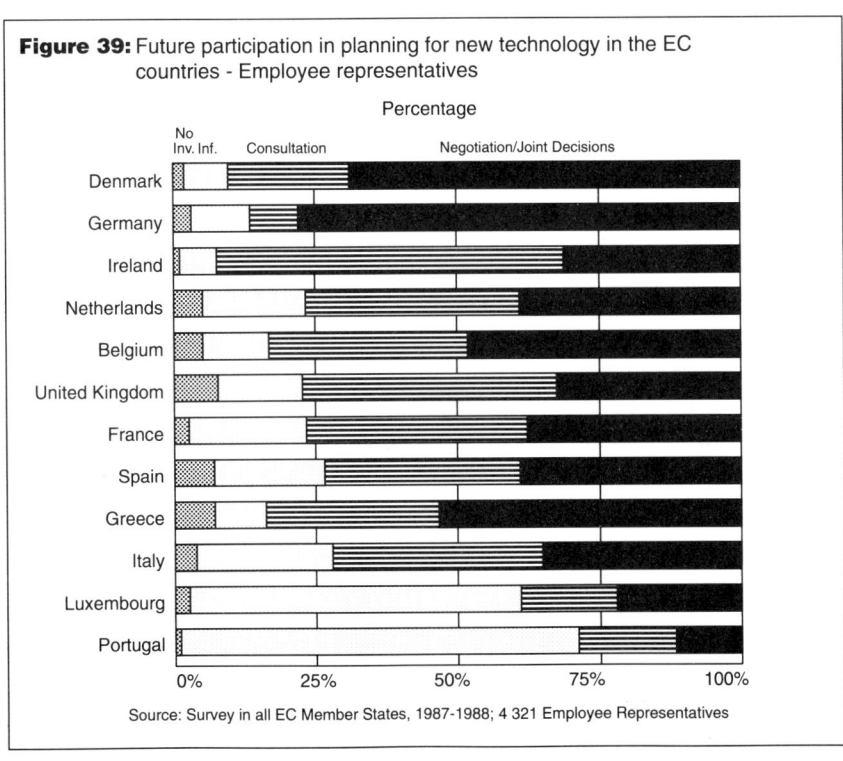

Source: Survey in all EC Member States, 1987-1988; 4 321 Employee Representatives

4. Do employee representatives support the policies of the ETUC in advocating negotiation or joint decision-making in both the planning and implementation stage of technological change?

5. How can our explanatory framework be used to explain the pattern of participation levels in each European Community member state?

1. DENMARK

The Danish system of industrial relations bears some resemblance to the Scandinavian 'model' of industrial relations in the sense that the employers' associations and trade unions are highly centralised, and both exercise almost complete formal power over their respective members. Both parties are well organised and cover almost the entire labour market. Denmark is characterised by a very high trade union density, which results from the historic role that the trade unions have had for administering unemployment benefit.

Pay and conditions of work have been regulated for many years through central 'Basic Agreements', which provide an explicit, formal normative framework for the settlement of both substantive and procedural matters through industry-wide agreements which are then applied after local bargaining to the individual establishments. The Danish central 'Basic Agreement' has served to provide a stabilising influence on Danish industrial relations throughout the century. In addition, legislation has established governmental agencies which are used in dispute settlement and the state may intervene in disputes of its own accord when certain conditions apply.

Danish industrial relations have been characterised for a considerable time by a well-established system of industrial democracy, which takes place at both board level and plant level. Danish company law provides for a unitary system of management boards elected by a general meeting of the shareholders. In 1974 two Companies Acts came into effect which provided for the participation of employees in management decision-making by offering them two seats on the boards of public and private limited liability companies which employed more than 35 workers. Recent legislation now enables employees to elect up to half the board members. The nomination and election of worker directors takes place outside union machinery, although in practice the senior shop steward in a company would normally have a seat on the board. Board representatives serve for four years and have the same rights and duties as the other board members. At first sight it might be thought that such measures would contribute to a redistribution of power in favour of employees in Danish enterprises, but a European Foundation Report concluded[2] that it had not, although the employers see the scheme as having been successful in promoting a much more favourable industrial relations climate.

At plant level industrial democracy in Denmark takes the form of **'co-operation committees'** in industrial and craft companies employing more than 35 people, a measure derived from a 1970 agreement (renewed in 1986) between the Danish Employers' Confederation (DA) and the major trade union confederation in Denmark, the LO. At national level there is a Co-operation Board consisting of three representatives from the employers' side, three from the trade unions, and one from the Foremans' Association. This body provides the guidance to industry concerning the implementation of the agreement, particularly in the field of advisory services to unions and employers to further 'co-operation' in Danish enterprises.

It is important to stress that the philosophy of co-operation committees is rooted in the notion that both parties should 'strive for agreement', with a commitment on both sides to ensure that the principles agreed are, in fact, applied.

There are more than 3000 Co-operation Committees in the private sector in Denmark. According to the Agreement, their primary task lies in observing and promoting day-to-day co-operation and involving as many people as possible in this task. The Co-operation Committee is restricted to discussing the principles of working conditions or work organisation; it may not discuss any matters relating to industry level collective agreements or company pay agreements which are dealt with through collective bargaining procedures or conciliation procedures.

How can we account for the survey evidence which shows that **Denmark has the greatest degree of worker participation in both the planning and implementation phases of new technology of all the 12 EC member states?**

In the **planning** phase of technological change **less than one in five Danish managers report that there had been "no involvement" by employee representatives — by far the lowest of all the countries in our survey.** Just under a quarter of employee representatives agreed that this was the case. 55% of managers interviewed stated that employee representatives were either consulted or were involved in negotiation or joint decision-making when technological change was being planned. Employee representatives were more sceptical in their interviews about the level of their involvement, but even if we take the opinions of employee representatives alone, just under a third of them are involved with management when major decisions about the form and direction of technological change are taken.

Figure 40: Past employee participation in planning and implementing new technology

Denmark

Planning: No Involvement | Information | Consultation | Negotiation/joint decision
- Managers
- Employee represent.

Implementation:
- Managers
- Employee represent.

Source: Survey in Denmark 1987:
Managers and Employee Representatives each N = 278 (weighted), N = 314 (unweighted)

Figure 41: Future employee participation in planning and implementing new technology

Denmark

Planning: No inv. | Information | Consultation | Negotiation/joint decision
- Managers
- Employee represent.

Implementation:
- Managers
- Employee represent.

Source: Survey in Denmark 1987:
Managers and Employee Representatives each N = 278 (weighted), N = 314 (unweighted)

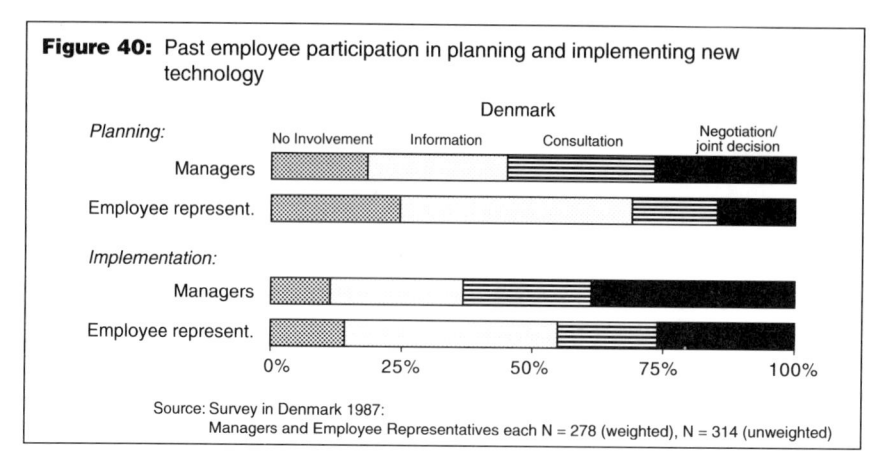

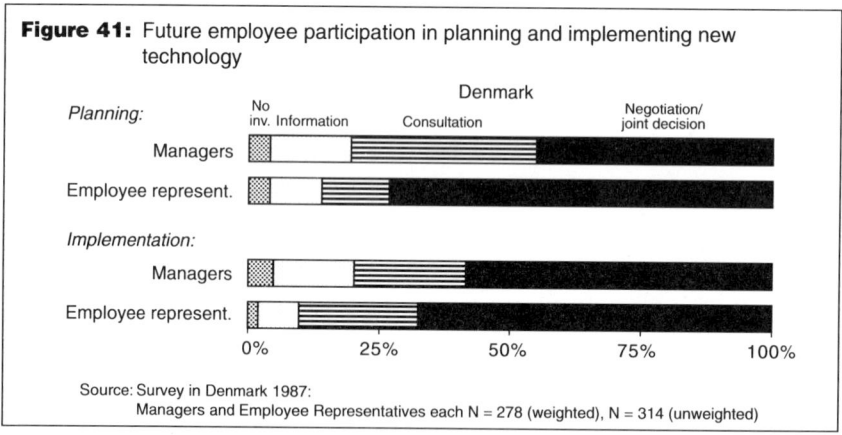

As we would expect, there is a greater degree of participation in the **implementation** phase, but only marginally so compared to other European Community states. 63% of Danish managers and 45% of employee representatives report that levels of participation ranged from consultation upwards.

When we look at the levels of participation that both parties envisage in the future, the **Danish results show a dramatic increase in the intensity of participation in both phases of technological change.** Four out of every five managers intend to either consult with their employee representatives or to tackle technological change by negotiation or joint decision-making in the **planning** phase — a very impressive result by European standards. Somewhat surprisingly, Danish managers do not differentiate between the planning and implementation phases; they **are equally enthusiastic to have participation levels of consultation or above in both phases.**

Similarly, Danish employee representatives do not appear to differentiate between the planning and implementation phases either. They are equally enthusiastic about their preferences for the future; 86% of them want levels of participation of consultation and above in the planning phase and 90% of them opt for similar levels of participation in the implementation stage.

One point stands out from Figure 41: **there is a great deal of disagreement between managers and employee representatives in Denmark about the kind of participation that they envisage for the future**. Employee representatives, are much keener than their management counterparts in seeking high levels of participation in technological change in the future; whereas only 46% of managers are willing to contemplate decision about technological change being made by either negotiation or joint decision-making in the planning stage, 73% of employee representatives feel that these forms of participation are best. There is not so much disagreement between the two sides in the implementation stage; here, 59% of managers prefer these forms of participation compared to 68% of their employee representative counterparts. Whether this disagreement between the two sides of industry in Denmark foreshadows conflict in the future is open to question, because although Denmark was always been noted for its conflict-free, stable industrial relations system with a high-trust relationship between management and employees, the incidence of disputes there has risen sharply since the mid-1980s.

Denmark is the only country in the European Community in our survey where a majority of managers state that the planning of technological change is characterised by consultation or higher participation levels. This claim, however, is contradicted by employee representatives; only 31% of them agree that this is the case. The Val Duchesse Social Dialogue Joint Opinion of the working party on 'Social dialogue and new technologies' concerning training and motivation, and information and consultation, which was concluded in 1987 by UNICE and the ETUC specified that there should be consultation or higher forms of participation when technological change was introduced into companies. If we take managers' responses to our survey as an indication of the pattern of participation by employee representatives in technological change in Denmark, this country is the only EC member state where a majority of managers are adhering to the Val Duchesse guidelines.

As to the future expectations of both parties, **four out of every five Danish managers are prepared to have participation levels of at least consultation or negotiation/joint decision-making in the**

planning stage in line with Val Duchesse. However, there is a great deal of disagreement about what is desired in the future between the two parties in the planning stage. Nearly **three out of every four employee representatives want this stage to involve decisions taken either by negotiation or jointly between both parties.** To what extent do employee representatives follow the policies of the European Trade Union Confederation? The ETUC maintain that the planning and implementation stages of technological change should be decided either by negotiation or joint decision-making. 73% of employee representatives adhered to ETUC policies in their preferences for the future in the planning stage of new technology and a smaller number (68%) in the implementation stage.

It is easy to appreciate why Denmark ranks the highest in our survey in terms of worker participation in the planning and implementation stages of technological change (see Table 8.1). The Danish industrial relations have a highly centralised system of collective bargaining, there is a high trade union density and the unions have a great deal of bargaining power. There is a duty on employers and employer representatives in Danish companies to provide information to the Co-operation Committee on all aspects affecting management-worker relations e.g. the company's financial position; future prospects; employment levels and planned organisational changes. The Co-operation Committees have consultation rights on general policies of day-to-day production and work planning as well as on the implementation of major changes in enterprises, particularly on new technology. The Co-operation Committees also exercise co-determination in formulating principles on work organisation, general personnel policy below supervisory level and the training and retraining of employees working with new technology. There are similar provisions for Co-operation Committees in both the public sector and the municipal sector. In the public sector the rules of co-determination at workplaces were adopted in 1972 and renewed in 1980. In the municipal sector, the negotiations leading to agreement on the rules of co-determination were completed in 1973 and revised in 1981.

In sum, all the factors which we identified in our explanatory framework in Chapter VII which shape the opportunities for employee participation in technological change are favourable (see Table 8.1 opposite)

TABLE 8.1

Variable

Technological Objectives	High dependence by management on the skills and problem-solving abilities of employees
Management Style	Co-operative and a willingness to negotiate over technological change
Bargaining Power	Highly unionised; trade unions cover the entire labour market and have a technically knowledgable and skilled membership; trade unions are well-resourced with well-trained officers and employee representatives; little inter-union rivalry
Regulation	Legislation which provides for the election of employees to the Boards of Danish companies; Established traditions of tri-partite corporatism
Industrial Relations System	Highly centralised; both employers and trade unions are able to exert authority over their respective affiliates; both parties cover the entire labour market

2. GERMANY

The German industrial relations system has a dual structure. At workplace and plant levels there is no direct bargaining between unions and employers. Instead, works councils and employers negotiate on a statutory basis. It is at industry-wide and regional level (and less often at enterprise levels) that unions and employers' federations enter into negotiations which usually result in collective agreements. There is a distinction between framework agreements, which usually last for some years, and 'ordinary agreements' which usually last for a year and regulate major conditions of work (pay, working conditions etc). Some of the ordinary agreements are concluded between one employer and the relevant union (enterprise agreements), the most well-known of which is perhaps that at Volkswagen[3].

West German unions are based upon an ideological pluralism, of which the social-democratic faction is the most influential although the Christian Democrat faction is instrumental in keeping the CDU relatively benign towards unions in general; radical factions have attained influence only at local levels. On the whole, union policy is highly centralised because of three factors: the industry-level collective bargaining system, the fact

Figure 42: Past employee participation in planning and implementing new technology

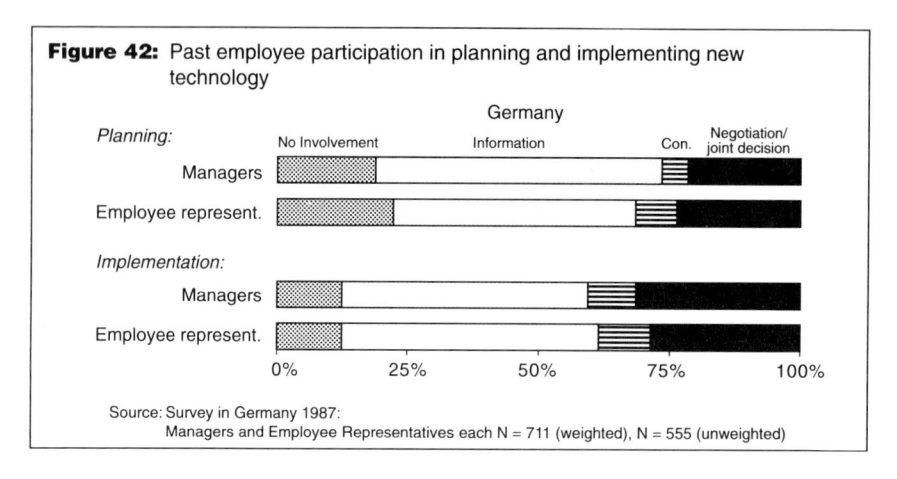

Germany

Planning categories: No Involvement | Information | Con. | Negotiation/joint decision

Planning:
- Managers
- Employee represent.

Implementation:
- Managers
- Employee represent.

Scale: 0% – 25% – 50% – 75% – 100%

Source: Survey in Germany 1987:
Managers and Employee Representatives each N = 711 (weighted), N = 555 (unweighted)

Figure 43: Future employee participation in planning and implementing new technology

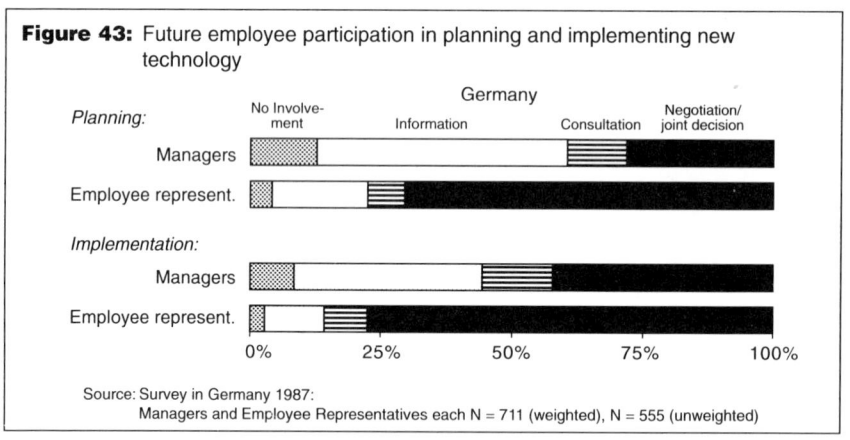

Germany

Planning categories: No Involvement | Information | Consultation | Negotiation/joint decision

Planning:
- Managers
- Employee represent.

Implementation:
- Managers
- Employee represent.

Scale: 0% – 25% – 50% – 75% – 100%

Source: Survey in Germany 1987:
Managers and Employee Representatives each N = 711 (weighted), N = 555 (unweighted)

that unions pursue policies which embrace wider social issues, and the highly bureaucratic and legalistic nature of industrial relations which requires the involvement of experts. Union policy extends to wider societal issues as well as the protection of members' rights in industry; for example, there is union concern about technology. Moreover, unions receive information and are consulted about all major areas of social and economic policy, including the quality of working life of the working population and their dependants. Although the unions are technically neutral in party politics, their political presence is substantial on economic and social issues.

There are factors evident in Germany which make technological change much less contentious than is the case in some other EC member states: management are actively committed to continuous technological innovation and are competent to initiate and establish new technical systems; there is an ample supply of all-round skilled labour which can be broadly deployed and easily retrained; there is a well-established training system and a willingness by both management and labour to invest in training and retraining and technological change which is accompanied by a relatively high degree of employment security and a low degree of labour market segmentation. Germany also provides an example of a co-operative system aiding worker identification with, and joint responsibility for, the efficient and competitive organisation of production.

Germany ranks second to Denmark in the survey results in having a high level of participation in both the planning and implementation phases of technological change. Taking managers' responses alone, 21% of German managers stated that participation in the planning phase was characterised by negotiation or joint decision-making and 31% stated that the **implementation** of technological change was carried out by negotiation or joint decision-making. 19% of German managers reported that there was no involvement in the **planning** phase and the percentage for no involvement in the **implementation** phase was 13%. The opinions of employee representatives hardly diverged from that of their management counterparts.

What form of participation do managers and employee representatives envisage for the future? The first point to note is **that there is a substantial divergence of views between both parties. Three out of five German managers either prefer no involvement or simply the impartation of information in the planning** phase of technological change. Employee representatives are much more hopeful; only 22% of them agree with this view.

When we consider the views of the parties in the **implementation** phase there is a similar difference of views. 56% of managers are prepared to see participation levels consisting of consultation or negotiation/joint decision-making, whereas **86% of German employee representatives want these forms of participation in technological change — similar to the views of employee representatives in Denmark.** As we mentioned in the case of **Denmark, this would suggest that there is much scope for conflict between management and employee representatives in the future about the form of participation in technological change.**

If we assess these results of both past and future opinions of both parties in the light of the Val Duchess UNICE/ETUC Joint Opinion, it is clear that large numbers of German managers are not conforming with the Val Duchess guidelines. However, the rank-and-file trade unionists of German unions show a high level of

support for the ETUC position on seeking negotiation or joint decision-making as the best way to deal with the planning and implementation of technological change.

Whilst Germany ranks high in participation in technological change relative to other EC member states, one must ask the questions why there is so much divergence between the views of both sides in their preferences for the future and why the German survey results do not show an even greater degree of participation in Germany than is the case? After all, Germany is noted for its extensive works council co-determination provisions, its "Humanisation of Work Programme" (Humanisierung der Arbeit: HdA) and the "Production Technology Programme" (Produktions — und Fertigungstechnik) of the late 1960s and the early 1970s. Coupled with this is the provision of State financial support for quality of working life initiatives at workplace level managed jointly by employers and Works Councils in a corporatist or co-decision relationship. In addition, the German trade unions have always sought to build up the effectiveness of workplace control over new technology ever since the mid-1960s and new technology has always been given a high policy priority.

One explanation which can be offered for the difference of views between both sides about the level of participation that is preferred in the future lies in the growing polarisation between employers and unions in Germany during the 1980s. The Conservative-Liberal government which has been in office since 1982 has pursued policies of de-regulation, sought greater flexibility in working practices, and has generally supported employers' demands for greater managerial authority in German industry. One effect of the compromise which emerged as a result of the 1984 strike over the shorter working week was that there was more authority given to works councils at plant level to implement the new agreement on shorter working hours. Despite the attempts of IG Metall to ensure that a standardised model on working hours was implemented throughout German factories, trade union members on works councils tended to negotiate different versions of the national agreement. As a result, differences emerged between the national union of IG Metall and trade union representatives on some works councils which encouraged the employers to pursue their drive for more flexibility and limited decentralisation.

It is worth comparing the German results with those of **Denmark.** One general reason for the lower level of participation in Germany is that German employers are vigorously opposed to any improvement of the legal provisions in this area — improvements which have been demanded by the DGB for a considerable time. The Scandinavian management style is considerably more open and as we saw in the case of Denmark, management there are willing to open up the subject of technological change to negotiation with the trade unions. One might speculate from this that they do so because collective agreements are seen by them as less permanent than is the case with legislation. Having said this, in both countries there is a considerable difference between management and employee representatives in both countries about the level of participation that they desire in the future. It seems that, given the comparatively high level of participation that exists in both countries, employee representatives want to build further on the provisions that they already have for the future.

One factor which limited the potency of German works councils in the 1980s was the increasing concentration of firms in West German industry[4]. The number of sole proprietorship firms is declining and the increasing number of multi-plant firms means that local management at plant level often have limited authority to take major decisions on technological change. The works council in a plant that belongs to a larger enterprise is faced with a management that has increasingly less authority and — as was proved during the implementation of the 38.5-hour week collective agreement — a management that, with regard to executive matters, is left with only limited freedom of decision-making in internal negotiations[5]. Moreover, the level in between the sphere of national collective bargaining (where the union is involved) and plant-level consultation and co-determination (where the works council is involved) there is a lack of representation on the employees' side. Relevant bodies for the representation of employee interests in the form of a group works council (Gesamtbetriebsrat) have been established by the legislature in the Works Council Act for enterprises and company groups, but they are provided with only little formal authority in comparison with the works councils at plant level; this, in reality, makes it impossible for the Gesamtbetriebsrat to force their local branches to adhere to particular policies against their will.

The new rights gained by workers under the 1972 Works Constitution Act (Betriebverfassungsgesetz) led to an expectation that their capacity to manage the processes of technological change would be substantially enhanced. Paragraph 90 of the Act provided for the works council to be informed "at an appropriate point" of changes in premises, technical equipment, production processes or working conditions. Such changes must be discussed "in the light of the accepted findings of scientific and engineering research in relevant areas". Paragraph 91 provided the right for works councils to prevent or to modify proposed changes in working conditions which are contrary to these accepted findings, provided that they are "significant".

Why then do we not find higher degrees of worker participation in technological change from the survey data in Germany? To answer this question it is important to appreciate the patterns of collective bargaining that exist in the German industrial relations system.

The German industrial relations system is characterised by a dual structure of representation, in which employee-based works councils are the sole channel for worker participation within the plant, and union-based collective bargaining at industry or regional level is responsible for general conditions of remuneration and employment. The strength of the system has arguably been its stability and predictability at industry-level and its capacity to give organised labour a powerful voice in the national corridors of power. But, for unions, the inability to bargain at workplace level except via the works council has resulted in a serious weakness. They have been largely unable to ensure effective co-ordination and integration of works council and national union policies; and the legal provisions covering works councils have placed awkward restrictions on the scope of their activities. In addition, the decline in manual jobs and their replacement by white-collar occupational categories coupled with the decline of the coal and steel industries with their "strong" forms of co-determination, are posing real problems for the DGB.

For example, union/works council influence is very weak in practice in the process of decision-making surrounding investment and development programmes; their real influence is confined to the area of "humanising" working conditions **after** key decisions about innovation and change have been made by management. Even where works councils are able to exercise their full information and consultation rights, they can only actually negotiate and exercise their co-decision rights over the social consequences of previously determined managerial decisions. There is, of course scope for considerable debate about what constitutes the definitions in Paras. 90 and 91 of the Works Constitution Act of "accepted findings" and the "appropriate time" for information disclosure. The ability to achieve significant concessions in plant negotiations/consultations is in any case affected by the peace obligation and the co-operation and confidentiality requirements of the Works Constitution Act, and also through council's limited ability to mobilise external support and resources.

Four specific difficulties can be identified for German works councils in obtaining adequate information about future plans in time to effectively influence them. Firstly, the limited definition of information rights results in certain types of information being legally withheld: leasing of equipment rather than direct investment, for instance, or small investments below a given value.

Secondly, employers can refuse to provide information on legal grounds and then continue to process these plans while works councils seek a legal resolution of the conflict; status quo does not exist. Thirdly, conflicts and communications difficulties between central and plant works councils are frequently a major obstacle. Finally, poor co-ordination between works councils and unions can lead to a reluctance to press for further information since there is no clear idea of how it might be used in employees' interests.

At the beginning of the 1980s the German trade unions hardened their attitude to what they saw as these inadequacies of the works council system and initiated a change in their policy on new technology. The clearest statement of this new approach can be found in IG Metall's Action Programme, which emphasises the importance of increased funding for Humanisierung der Arbeit programmes, an employee-orientated industrial policy of reducing working-time and improving working conditions, to open up works council activities to influence and support from external bodies and to press the government for an extension of co-determination rights. This change in policy orientation was reflected in the increasingly aggressive campaign to reduce the working week as a means of creating new jobs and distributing some of the economic benefits of technological progress towards workers. The campaign was spearheaded by IG Metall and IG Druck und Papier, but despite the lengthy strikes in the summer of 1984, results have not met the unions' expectations. Nevertheless, the issue of working time has become a clear element in German unions' bargaining strategy, linking the "quantitative" objective of improving benefits, with the "qualitative" desire to counteract what they see as the employment consequences of rationalisation.

The attempt on the part of the German unions to open up works council activities to influence and support from external bodies builds on the experience gained since 1979 in the work of Technology Advice Centres (Technologieberatungsstellen), established in Oberhausen by the DGB, and in Hamburg and Berlin by IG Metall. These centres provide support and advice to works councils on the introduction of new technologies, and train works council members and union representatives on new technology issues. They provide information, assess management proposals and have become actively involved in the development of alternative "worker plans". In addition, the Economic and Social Research Institute (Wirtschafts und Sozialwissenschaftsciches Institut — WSI) of the DGB, the Hans-Böckler Foundation (DGB) and the IG Metall have information and advisory departments covering new technologies.

These information and support services have undoubtedly been of considerable value to works councils in local negotiations, but a great deal more needs to be done if the "quality of working life" objectives are to be achieved. As a key part of its action programme for Work and Technology, IG Metall is intending to strengthen the training and advisory services provided by its district offices; more local officers will be trained to act as "technology experts", able to both advise and to take independent initiatives, and local links with universities through co-operation agreements will be strengthened.

The new stress on the plant as the key focus of the campaign, and the importance placed on the unions becoming actively involved in co-ordinating plant negotiations underlines the union's recognition that **their lack of direct impact at plant level** has hitherto been a major source of weakness in making an effective challenge to employer policies on new technology. Union leaderships have been able, at a programmatic level, to identify the collective bargaining pre-conditions for the introduction of new technology on an acceptable basis but they have been unable to ensure a generally effective translation of these programmes into action at company or plant level.

The union objective of using co-determination rights to control the introduction of new technology has given way in practice to the protection of existing workers' interests, with only limited attention being given to the design and control of new systems. At industry level, the employers have resisted the unions' claims for framework agreements on qualitative issues, and the unions have been unable to mobilise their membership around such abstract issues. Thus, the limited ability of the German unions effectively to implement their policies of controlling job and systems design is a **reflection of the "dualism" built into the works council/collective bargaining distinction.**

In summary, whilst Germany has a higher degree of worker participation in new technology than most of the other EC member states in our survey, further participation is limited by the absence of an effective union presence at plant level. Germany has five favourable factors which were identified as conducive to participation in technological change which were identified at the beginning of Chapter VII (see Table 8.2 opposite). However, its legislative provisions on the "dualism" built into the works council/collective bargaining distinction places a limit on the level of employee participation in technological change.

TABLE 8.2

Variable

Technological Objectives	High dependence by management on the skills and problem-solving abilities of employees; extensive provisions for training and re-training; labour force which can be easily deployed; active management commitment to invest in technological change
Management Style	Diverse but generally co-operative, particularly in larger companies but tends to be paternalistic in smaller enterprises; well-established works council system but its role in enabling employees to influence technical change is limited by the absence of a trade union negotiating role at workplace level; management vigorously opposed to further improvement in levels of participation through legislation
Bargaining Power	Moderate trade union density but influential central confederation in DGB; absence of trade union bargaining role at the workplace; technically knowledgable and skilled membership; trade union officers and employee representatives are well-resourced and experienced; trade unions are active in educating their members on technological change; concern for quality of working life; trade unions adopt a co-operative stance with management but seek balance of power in their favour through "real parity" co-determination
Regulation	Extensive legislation on works councils; collective agreements are limited to industry-wide level and do not cover technological change (apart from working hours reductions); some evidence of weakening of employee influence on works councils during the 1980s
Industrial Relations System	Fairly centralised at sectoral level; co-operative with low strike incidence; legalistic; little government intervention in industrial relations

3. IRELAND

The Irish system of industrial relations bears a close resemblance to that of the **United Kingdom**, although in the last two decades the two systems have moved much further apart. Despite the differences between the two systems it is still accurate to describe Ireland as retaining traces of "voluntarism" as a characteristic feature of its industrial relations. Like Britain, Ireland provides a system of "negative immunities" as a form of protection for trade unions from criminal or civil liability rather than "positive rights" which are often constitutionally based. The law has played only a minor role in terms of the origin, operation and outcome of collective bargaining in Ireland.

Since Ireland gained independence from Britain in 1922 it has developed its own industrial relations laws. Trade unions must gain a negotiating licence issued by the Ministry of Labour under the Trade Union Act of 1941. Disputes are settled by a tri-partite Labour Court which was established by the 1946 Industrial Relations Act and which provides a conciliation service; although the Court's recommendations are not legally binding they do have a strong persuasive influence on the parties to the dispute. The 1946 Act was amended in 1969 to provide for the appointment of Rights Commissioners. These Commissioners investigate on request by a party to a dispute — unless there is an objection by the other party — matters other than pay, hours of work and holidays. In practice, this means that they deal with dismissals and suspensions, and the Commissioners are able to make recommendations on the merits of the dispute in question. A party may appeal against the recommendation to the Labour Court, which hears the appeal in private and whose decision is legally binding.

As in Britain, collective bargaining takes place at national, company and plant level and varies from industry to industry and across occupations. Company bargaining tends to be particularly important but there are national Joint Industrial Councils in construction, electrical contracting, banking and baking. Where unionisation is weak and pay is low there are a number of tri-partite Joint Labour Committees comprising employers', trade unions' and independent representatives. These committees cover some 45,000 workers in areas such as tailoring, hairdressing, hotels and contract cleaning. The findings of these Committees are confirmed by the Labour Court and are then legally binding in the industry concerned; in effect they provide a quasi-form of minimum wage.

The trade union movement in Ireland — both in the Republic and in Northern Ireland — is united into one centre, the Irish Congress of Trade Unions (ICTU) although trade union organisation in Ireland is further complicated by the fact that the British TUC also operates in Northern Ireland. As in Britain, the trade union movement is characterised by a bewildering mixture of different types of unions which are organised across a wide range of industries and occupations, and there is much overlapping and inter-linking between them. There is a great deal of inter-union competition and rivalry, and despite the efforts of the ICTU, the amalgamations that have occurred among unions in Ireland owe as much to political expediency as they do to industrial logic. One peculiarity is that ICTU-affiliated unions include both unions based solely in the Irish Republic and Irish branches of British unions. Trade union density is around 45%.

Just as in Britain, organisations of shop stewards exist at workplace level, particularly in large companies in some industries. Their role varies from one establishment to another but they do fulfil an important function in representing the interests of trade union members at shop-floor level. There is no compulsion on the part of employers to recognise trade unions or to bargain with shop stewards in their companies and the influence of stewards varies substantially, depending on the industry or company concerned. The rights accruing to shop stewards derive from voluntary agreements between employers and trade unions.

Our survey data shows that the extent of participation by Irish employee representatives in the planning stage is almost identical to the overall involvement for the other European Community countries. Around a third of Irish managers reported that there was "no involvement" by employee representatives in the planning stage of new technology introduction whereas just over a half of all employee representatives had the same opinion. There was a similar difference of opinion between managers and employee representatives as far as "information/consultation" was concerned; 54% of managers claimed that this form of participation existed whereas this opinion was only shared by 39% of employee representatives. 13% of managers stated that employee involvement took the form of "negotiation" or "joint decision-making" compared with 9% of employee representatives. **The Irish results are broadly in line with the general European average for the planning stage of technological change.**

In the implementation stage the results for Ireland are slightly above the European average. There was a greater convergence of opinion between managers and employee representatives in this phase. 16% of managers reported that there was "no involvement" in this phase compared with 11% of employee representatives; nearly three out of five managers stated that technological change was characterised by "information/consultation" compared to 67% of employee

Figure 44: Past employee participation in planning and implementing new technology

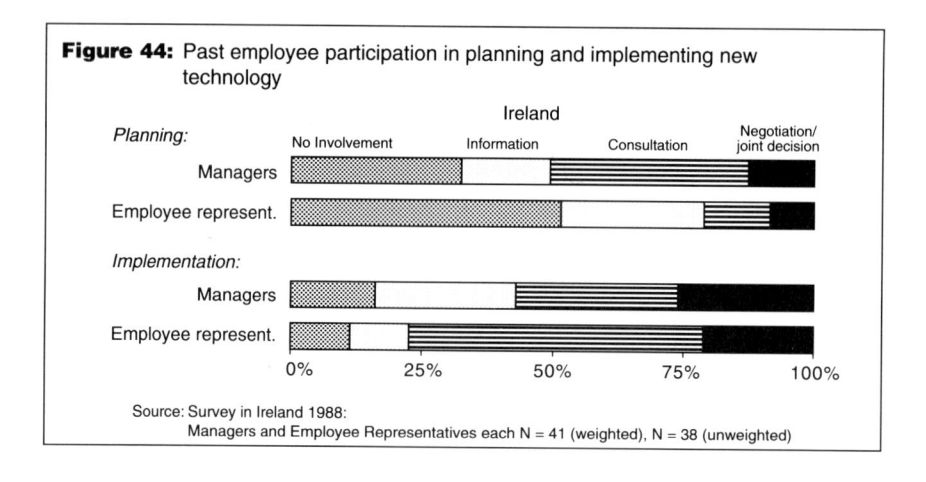

Source: Survey in Ireland 1988:
Managers and Employee Representatives each N = 41 (weighted), N = 38 (unweighted)

Figure 45: Future employee participation in planning and implementing new technology

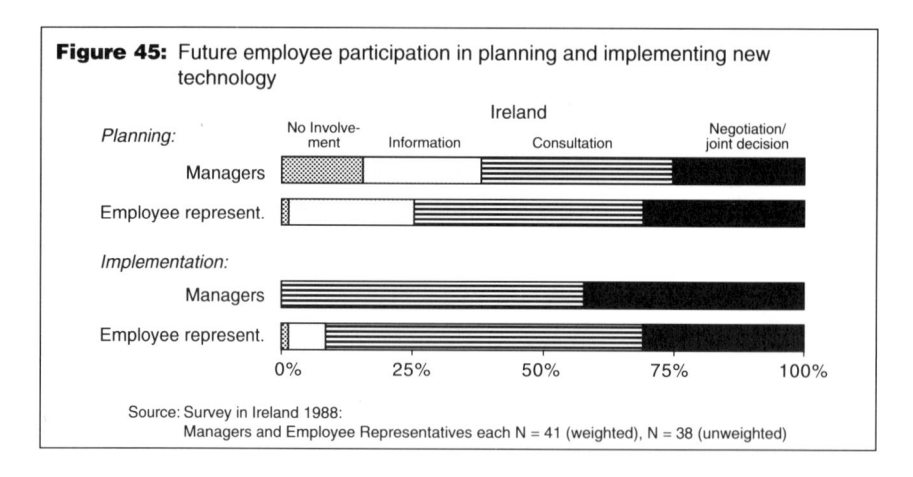

Source: Survey in Ireland 1988:
Managers and Employee Representatives each N = 41 (weighted), N = 38 (unweighted)

representatives. Just over a quarter of Irish managers described employee involvement as "negotiation" or "joint decision-making" compared to two out of five employee representatives. **Ireland thus ranks slightly above the European average in the implementation phase.**

It is important to emphasise that the Irish results should be treated with a great deal of caution. The number of respondents interviewed was very low compared to the size of the sample for other European Community countries so that, even allowing for the weighting procedure in our methodology, too much significance should not be attached to the results. One other problem of the Irish sample of respondents is that the sample was compiled from the Dunn and Bradsheet list of the top 500 Irish companies; in Ireland this directory is regarded by researchers as a poor source of information for survey purposes.

Another point to emphasise is that our survey was restricted to companies in industrial sectors which were not covered by the legislation enacted in 1977 associated with the Worker Participation (State Enterprises) Act. This piece of legislation provided for the election of worker directors in seven State companies (Road and Rail Transport, Shipping, Fertiliser manufacture, Electricity Supply, Peat Production, Air Transport and Sugar manufacture). The main features of this Act were as follows[6]:

One third of the Board membership to be elected by and from the workforce in the context of the Irish unitary board system.

Employees to be appointed directors on the basis of an election conducted by secret ballot under the proportional representational system, with one electoral list comprising the total number of employees eligible to vote in each company.

Universal suffrage to apply in each election, but entitlement to nominate candidates to be restricted to trade unions and staff associations recognised for the purpose of collective bargaining within the company.

Elected directors to hold office for three years with an entitlement to seek re-election, to have the same rights, responsibilities and duties as other directors on the board, and to be governed by the same regulations concerning directors' interests.

It could be argued therefore, that our results clearly underestimate the degree of worker involvement in decisions on technological change in the case of Ireland. Since 1977 the legislative provisions have been extended to telecommunications, postal services, forestry and manpower training, but again, these are not sectors covered by our survey data. There is much evidence from the Irish literature on industrial relations that Irish employers have always viewed legislative forms of worker participation with some hostility.

In the private sector most forms of participation in technological change are through consultative arrangements at shop-floor level through the shop steward system. Much of this consultation is of an informal kind and in the larger enterprises consultative committees also embrace a degree of collective bargaining. Depending on agreements and custom and practice, shop stewards, in addition to negotiating aspects of pay and conditions of employment, undertake liaison with management over production, productivity and manning questions and also represent their members in grievance and disciplinary matters.

Irish industry is characterised by a smaller average size of establishment than is the case in some of the major countries of the European Community. This has meant that there are not many formalised forms of participation in the private sector. The Irish Advisory Committee on

Worker Participation estimated in 1986 that there were only 50 works councils, or their equivalent, in existence in the Irish private sector. Throughout the 1980s there has been an increasing polarisation of attitudes between employers and trade unions on the issue of employee involvement in firms. According to Wallace[7], the Minority Report by the Federated Union of Employers (FUE) — issued as part of their involvement in the Advisory Committee on Worker Participation — opposed the idea of any "enabling legislation being introduced in the private sector...to give status and impetus to the development of participation" (Committee Report, 1986). The FUE argued instead for a voluntary approach geared towards the needs of individual companies and their employees.

The Labour Management Service of the Irish Productivity Centre (IPC), a body jointly sponsored and controlled by employers and trade unions and partly funded by government, does carry out a limited degree of promotional activity within Irish companies focusing on the need to develop and monitor forms of co-operation and participation through work organisation and representative systems. However, the resources available to the Labour Management Service of the IPC have not permitted it to service a wide network of experimental activity embracing different levels in different industries and locations.

There are some new technology agreements in Irish associated banks which were introduced following negotiation and agreement between the Irish Bank Officials' Association and the banking employers. There are also a large number of new technology agreements in Irish manufacturing industry which were concluded during the 1980s.

The Irish results show that both managers and employee representatives express a desire for greater levels of participation than has existed in the past. This is true for both the planning and implementation phases.

Well over half of all managers (54%) indicated that involvement was characterised by "information/consultation" in the planning phase but this percentage increases to 58% for their preference for the future. Similarly, four out of every ten employee representatives indicated that involvement was characterised by "information/consultation" in the planning phase, but again nearly seven out of ten would prefer this form of involvement for the future. 13% of managers indicated that past involvement in the planning phase was "negotiation/joint decision-making", whereas a surprisingly high number of them (over a quarter) favour this form of involvement for the future. Employee representatives showed an even greater shift in their preference for this type of involvement (one in ten in the past to three in ten for the future).

Irish managers show a surprisingly high preference for higher levels of employee representative involvement in the implementation phase. Over a quarter of managers (26%) stated that past involvement was characterised by "negotiation" or "joint decision-making" and more of them would prefer this form of involvement for the future (43%). This percentage is second only to Denmark of all the European Community countries. The desired level of "information" or "consultation" involvement shows little difference from past practice (58% in the past; 57% in the future). Strangely, a lower percentage of employee representatives would prefer "negotiation" or "joint decision-making" in the future than managers (21% in the past; 31% in the future).

It is clear that the Irish results compare very favourably with other European Community countries. However, one has to add the reservation to these results by allowing for two caveats: the small sample of respondents who were interviewed in the survey and the fact that the industrial sectors in which the survey took place were not covered by the worker participation legislation.

What are the prospects for increased levels of participation in Ireland in the future? There have been a number of "National Understandings" between Irish employers, the ICTU and the Government in the 1970s and the late 1980s on industrial relations. Nevertheless, the Irish employers have accused the unions of using the worker participation issue in a cynical way, shying away from consideration of works councils and job enrichment and opting for directors on the principle that this policy demands least from the trade unions and most from the employers. The employers' reluctance to embark on schemes which would extend the worker director principle to the private sector is defended, therefore, on the grounds that the trade unions have little enthusiasm for any departure from their traditional role of relying heavily on collective bargaining.

There are a number of constraints which limit the prospects for greater levels of participation by Irish employees in technological change; the low levels of management and supervisory development; high levels of distrust between management and unions and limited access to information; the multiplicity of trade unions and the rivalry that exists between them; and the absence of flexibility between craft and general functions in manufacturing. In addition, trade union representatives are ambivalent about consultations with management and the acceptance of joint responsibility outside a bargaining situation. As for management, there is frequently apprehension about handing over power without any assurance of assistance for management's task. Crucial to any extension of employee participation in technological change is the disclosure of information. It would appear that the employers favour a voluntary approach that is mutually agreed between the parties, while the trade unions, contrary to their usual position, seem to favour a legislative approach. The Irish government is increasingly moving towards a preference for legality.

How do Irish managers view employee involvement in technological change in the light of the Val Duchesse joint opinion between the ETUC and UNICE? Half of all managers state that employee involvement in technological change in the planning stage and 43% in the implementation stage amounts to no more than information or no involvement. As for their future preferences, there is little indication of a change of attitude as regards the planning stage; four out of ten of them do not envisage a level of involvement greater than simply providing information or having no involvement at all. However, there is a much greater shift in opinion for the implementation stage; over four out of five managers are willing to envisage employee involvement at levels greater than simply providing information. Indeed, not one manager interviewed opted for no involvement.

Generally speaking, our survey results for Ireland show that it features in the top half of the European table for levels of participation. There are a number of favourable factors which contribute to this, even allowing for the relatively small sample of respondents in that country. Compared with the United Kingdom, Ireland has a more favourable climate for participation.

Three out of our five explanatory factors are generally favourable in the case of Ireland as can be seen from Table 8.3 below:

TABLE 8.3

Variable

Technological Objectives	Little dependence by management on the skills and problem-solving abilities of employees; enterprises are generally smaller than the European average; low level of management and supervisory development
Management Style	Generally opposed to participation schemes but employers are prepared to take part in "national understandings" with the trade unions and government; few works councils in the private sector
Bargaining Power	Trade unions have significant bargaining power in certain sectors; inter-union rivalry and pockets of multi-unionism; trade unions not very well-resourced

Regulation	No legislation providing for participation in the private sector but recent "national understandings" between trade unions, employers and the government; heavy reliance on collective bargaining; some new technology agreements have been negotiated
Industrial Relations System	Generally decentralised, but the Irish trade unions are gradually moving towards a greater degree of centralisation; during the 1970s and the late 1980s there have been a number of "National Understandings" between employers, trade unions and the Government.

4. NETHERLANDS

The Dutch labour market is in many ways similar to that of the Scandinavian countries. However, it has suffered more severely than those countries from economic difficulties and during the 1980s the unemployment rate has risen. There are two major trade union confederations within the Dutch industrial relations system: the Christelijk Nationaal Vakverbond (CNV) with 16 affiliated unions and the Federatie Nederlandse Vakveweging (FNV) with 17 affiliated trade unions. Trade union membership in the Netherlands was hit particularly hard in the first half of the 1980s. In March 1980 trade union density stood at 39% and had decreased to 29% by March 1985. Apart from unemployment, there are a number of explanations for this steep decline in trade union membership: the shift towards the services sector from manufacturing; the decline in the average size of establishment; decreased employment concentration; the growth of flexibility in jobs; the increasing number of part-time jobs — most of which are taken up by women who are less likely to take up union membership; socio-cultural changes — the decline in the "male bread-winner" family unit and the growth of individualism; new technology; the existence of several trade union confederations in the Netherlands and cuts in public expenditure and public sector jobs.

Pay and other terms of employment are normally regulated by collective agreements at industry level between unions and employers' associations. Collective agreements cover about 60 percent of the labour market. Although the negotiations are handled independently, the government has considerable authority to intervene; all collective agreements must be approved by the Ministry of Social Affairs, which has had the right since 1970 to decide on a pay freeze for a given period if the economic situation merits such action. This right, which has been exercised several times, has often aroused a great deal of opposition from trade unions.

A Social and Economic Council or Social Economische Raad (SER) has considerable influence over economic and social issues. The Council, which includes representatives of the government, employers and employees, also takes up many other issues, including the application and diffusion of new technology.

In our survey the Netherlands results show that both managers and employee representatives indicate a high degree of intensity of participation in their companies compared to most other European Community countries. One in four Dutch managers state that there is "no involvement" by employee representatives in the planning stage of technological change and 15% report "no involvement" in the implementation stage. Both these percentages of "no involvement" are below the European average. If we assess the Netherlands results in the light of the Val Duchesse guidelines we can see that by taking managers' responses alone, just under a third state that there are levels of participation involving consultation and higher levels in the planning stage of technological change and nine out of 20 Dutch managers claim such levels of participation in the implementation stage.

If we use our explanatory framework to account for the comparatively high level of past participation in technological change in the Netherlands, there are at least three and possibly four factors which are favourable in shaping the opportunities for involvement by employees in technological change. There is a well-established tradition of management-employee dealings through works councils; it seems that Dutch management generally rarely seeks confrontation with the trade unions (the Netherlands has a very low incidence of strike-proneness by European standards); the Dutch government has demonstrated its awareness of the possible adverse effects of new technology on workers and the Dutch unions are relatively well organised on shopfloor level in large enterprises by European standards with an overall density of between 25-30%. Both managers and employees are well-trained by European standards in utilising new technology, and there is consequently a high dependence by management on the skills and co-operation of employees.

According to a survey carried out by the Dienst Collectieve Arbeidsvoorwarden (DCA),[8] which monitors collective agreements for the Ministry of Social Affairs, around 1.3 million Dutch workers are covered by agreements which contain clauses relating to new technology. 85 agreements were surveyed, covering 2.3 million employees, or around 80% of all employees covered by agreements.

Of these, 58 contained clauses referring to new technology in general, to automation, to specific production processes or to investment in new technology.

Figure 46: Past employee participation in planning and implementing new technology

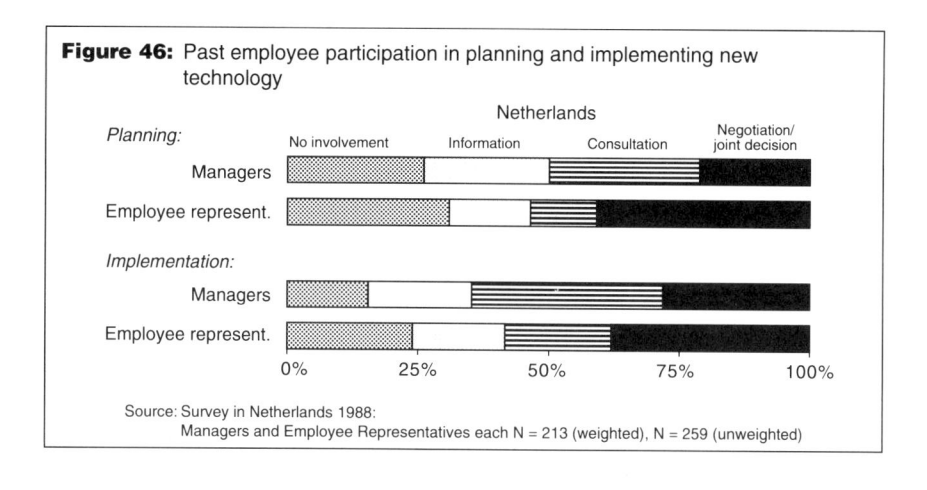

Netherlands

Planning: No involvement | Information | Consultation | Negotiation/joint decision
- Managers
- Employee represent.

Implementation:
- Managers
- Employee represent.

0%   25%   50%   75%   100%

Source: Survey in Netherlands 1988:
Managers and Employee Representatives each N = 213 (weighted), N = 259 (unweighted)

Figure 47: Future employee participation in planning and implementing new technology

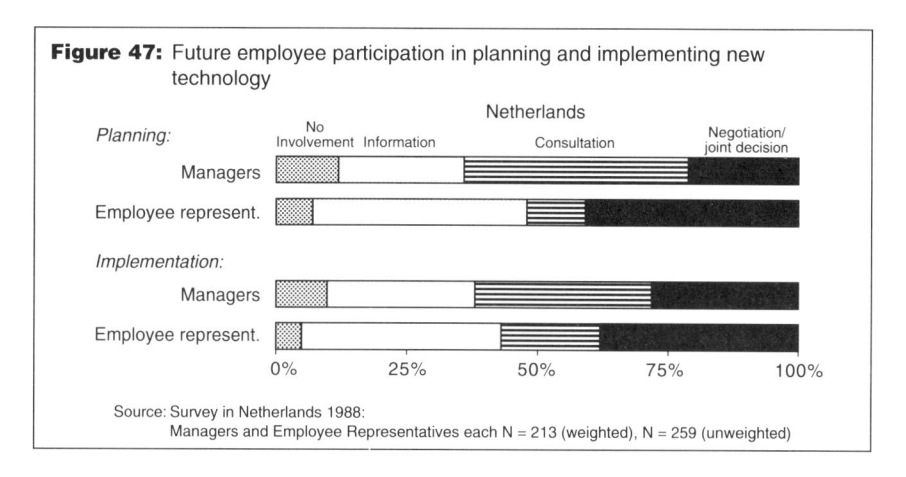

Netherlands

Planning: No Involvement | Information | Consultation | Negotiation/joint decision
- Managers
- Employee represent.

Implementation:
- Managers
- Employee represent.

0%   25%   50%   75%   100%

Source: Survey in Netherlands 1988:
Managers and Employee Representatives each N = 213 (weighted), N = 259 (unweighted)

The most common clauses — in 50 agreements covering 1.1 million workers — concern the right of employees to be informed and unions and works councils to be consulted over proposed new technology and its implications for work content and employment.

In fact, the main channel for possible participation by employee representatives in the Netherlands is through the works council system. The main legislation governing the setting up and operation is the Works Council Act, 1979 (Wet op de Ondernemingsraden); this replaced earlier legislation and gave works councils greater rights and powers. An amendment passed in 1982 extended a number of these rights and powers to companies with between 35 and 100 employees, which are now also required to establish a works council. The 1982 amendment also provided that companies with between 10 and 34 employees may, if they wish, set up a works council but they are not obliged to do so. They must, however, hold meetings of the management and employees at least twice a year, or more often if so requested by at least 25% of the workforce.

The legislation provides for works councils to have rights to information and consultation on a number of areas relating to the company's operations and the interests of the employees, but it does not give them powers of negotiation on pay and other terms and conditions. They must, by law, meet the employer at least six times a year and have the right to obtain such information about the company as is necessary for their proper operation.

They may set up special sub-committees and have the right of advice from outside experts when necessary. Works councils may meet during working hours, at the workplace and also have access to facilities (e.g. accommodation, telephone etc.).

Works councils in the Netherlands are composed entirely of employee representatives and their number increases according to the size of the undertaking. All employees, except those working few hours or 'on loan' from other companies, are entitled to vote and to stand as candidates in biennial elections. Although voting and candidature is not limited to union members, 65-70 per cent of the counsellors belong to FNV, CNV and MHP unions, more than double the size of these unions among the electorate.[9] While far from being part of, or dependent upon, the larger union organisation, the councils have become the centre of worker organisation in firms.

Under the 1979 Act, works councils have a large array of legally specified consultation and co-determination rights. Management must seek the council's advice with respect to major commercial decisions, including mergers, closures, takeovers, major investments and loans, hiring outside consultants and the employment of temporary staff. It is required to seek the council's approval in matters of changes in renumeration, pension and profit-sharing schemes, the arrangement of hours and holidays, grievance procedures, health and safety conditions and training. They have three main legal rights: to initiate discussions with the employer on specific issues; to be consulted by the employer and to approve certain decisions.

How do these bodies function in practice? A report which was commissioned by the Dutch Ministry of Social Affairs and Employment in the mid-1980s and carried out by the Institute of Applied Sociology in Nijmegen[10] covering a sample of 173 works councils concluded that while such bodies had become part of the industrial relations landscape in the Netherlands, there was evidence to suggest that they had not used the 1979 legislation to increase substantially their share in decision-making at company level. Another research study concluded that the formal agenda of new technology does not always lend itself to negotiation and is often difficult for works councils to negotiate.[11]

On the right that they have to initiate discussions on a specific topic with the employer — seen as the least important of the three main legal rights by employee representatives — the employers and the employee representatives gave different assessments of how often this right was used. The employers claimed it had been used more frequently than did works council chairmen. Two thirds of these works council initiatives were on personnel issues and in about three quarters of all cases the company accepted wholly or in part the proposals made by the works council. Where such an initiative is taken the employer is under an obligation to delay any action until the discussion takes place.

A more important right of Dutch works councils is the right to be consulted and to give their opinion on various major issues affecting employees in the company. If the works council does not give a favourable opinion, the employer must delay implementing his proposal for one month during which time the works council may refer the matter to a special court, the "ondernemingskamer", which has powers to require the employer to withdraw the disputed proposal.

The Nijmegen survey found that while there was disagreement between management and works councils about how the law should be interpreted, requests for an opinion on major organisational issues nearly always took place, although in 68% of all cases such requests came after a decision had in principle been taken and in only 32% of cases before a decision was taken. Nevertheless, in almost three quarters of cases works councils gave a favourable opinion on the proposal put forward by management. According to the survey, works council chairmen do not feel they have any great power to change the terms of a management proposal: only 7% of respondents said that they could always change a proposal and 37% said that they could sometimes change it. Of course, this might be because managers took into account the possible objections of the works council before making a proposal. Reference of such matters to the special court rarely took place. Importantly, more than half the works council chairmen felt they needed better access to information in order to give their opinions.

The third legal right that Dutch works councils have is to approve certain decisions taken by the employer (instemmingsrecht) which mainly cover a range of personnel matters. If a decision is withheld, then the matter is referred to a bi-partite industrial commission from which the employer must obtain permission. Works councils gave immediate approval in 59% of all cases and after some modifications in virtually all other cases. In only 4% of cases did an employer proceed with action which had not been approved by the works council.

It appears that employees are not all that keen to become members of works councils; 30% of works councils reported difficulty in filling vacancies. The job of a works council chairman appears to be somewhat isolated; 86% of chairmen surveyed felt their works council received little input from employees. The links between trade unions appear to be tenuous; fewer than one third of all councils had regular contact with trade union officials although about two thirds of all members were trade unionists.

Another factor which is favourable in assisting Dutch employees to participate in technological change is the role of the government. A report presented by the Minister of Social Affairs and Employment to the lower house of the Dutch Parliament in 1987 on the social aspects of new technology stressed that the field is one for joint regulation between the two sides of industry.

Nevertheless the government does help Dutch workers by subsidising courses for trade union officials and gives grants to the main trade union confederations to enable them to develop courses specifically on the subject of new technology; a total of 3.6 million guilders was allocated in the 1987 Budget to the main confederations FNV and CNV for this purpose. The government also funds courses for members of works councils, in particular to enable regional works council centres to be better equipped to help individual works councils with enquiries and specific problems.[12]

Finally, the introduction of the Working Conditions Act (Arbo-Wet) must be mentioned. The so-called Arbo Committee on Health, Safety and Welfare is also dealing with the quality of working life.

As to job satisfaction, consultation procedures are slightly more prominent, and in regard to health and safety, co-determination is the participation measure most often exercised.

Our survey data shows that the two sides of industry in the Netherlands would prefer more participation in technological change in the future. If we consider the views of managers with respect to the planning stage the number of those favouring "no involvement" by employee representatives in the future halves (26% to 12%); the percentage of employee representatives decreases even further from 31% to 6%. If we assess these preferences for the future in the light of the Val Duchesse Joint Opinion guidelines, we see that 65% of Dutch managers are prepared to entertain levels of participation involving consultation and stronger forms in the planning stage of technological change — second only to Denmark and well above Germany. Although employee representatives would prefer higher levels of participation in planning of technological innovation in the future, there is not a great deal of disagreement between the two sides. In the implementation stage, the percentage of managers who are prepared to fulfil Val Duchesse guidelines is even higher (71%).

How many Dutch employee representatives follow ETUC policies of insisting on technological change being introduced in both the planning and implementation stage by negotiation or joint decision-making? Just over four out of ten want such forms of participation in the planning stage (the third highest of all the 12 member states), and a slightly smaller number favour ETUC policies in the implementation stage.

Overall then, the Netherlands ranks fairly high in employee participation in technological change compared to the data for other European Community countries. There is a great deal of similarity between the Dutch works council system and that of Germany, and one might legitimately ask why participation levels in the Netherlands in our survey are lower than in Germany. This might be because while the Dutch works councils have some influence over the composition of supervisory company boards, the Dutch legislation never followed the German example of extensive representation on such boards.

There are several favourable factors at work in promoting participation in Dutch enterprises, particularly the works council system which despite its limitations, seems to work reasonably well. We must qualify this by noting that about three in ten Dutch managers are reluctant to offer anything more than "information" in the implementation stage of technological change — below that envisaged by the Val Duchesse joint opinion. Just under one in four employee representatives in the Netherlands similarly opt for no more than "information".

The Netherlands results can be summarised by using our explanatory factors in Table 8.4 below:

TABLE 8.4

Variable

Technological Objectives	High dependence by management on the skills and problem-solving abilities of employees
Management Style	Generally co-operative; a large number of agreements contain a clause covering technological change; well-established works council system
Bargaining Power	Trade union density declining but still relatively strong; links with works councils somewhat tenuous; technically knowledgable and skilled

	membership; trade unions are well-resourced with well-trained officers and employee representatives; some inter-union rivalry between the FNV and CNV; government support for trade union research into new technology and its work implications
Regulation	Legislation which provides for works councils; many collective agreements provide for implications of technological change
Industrial Relations System	Fairly centralised but employers increasingly pressing for more decentralisation and flexibility; low strike incidence; extensive government powers to intervene in collective bargaining; concern for quality of working life by government

5. BELGIUM

Belgium is unusual among EC member states in that a national technology agreement was negotiated in December 1983 between the unions' and employers' representatives covering all enterprises in the private sector of the economy. Our survey data therefore provide a means of evaluating how the national technology agreement has worked out in practice.

In fact, the Belgian industrial relations system provides a very favourable context for the use of a national technology agreement. Thus an evaluation of the operation of the Belgian case could provide important information on the possibility for the use of technology agreements in other EC member states.

What then are the characteristics of the Belgian industrial relations system which provide such a favourable context for a national technology agreement? Belgian industrial relations is characterised by three fundamental principles which, taken together, have been called the "social compromise": there is a strict division between negotiable and non-negotiable matters, with wage increases, personnel problems, holiday entitlements etc. considered the subject of negotiation while investment decisions and the rationalisation of work being subject to managerial prerogative; there is a trade-off between economic growth and social progress whereby the former implies a distribution of gains between management and labour; finally, there is an emphasis on "industrial peace", with agreements made between trade union officials and employers' associations being translated into operation at company or plant level.

Figure 48: Past employee participation in planning and implementing new technology

Belgium

Planning: No Involvement | Information | Consultation | Neg./j dec.
- Managers
- Employee represent.

Implementation:
- Managers
- Employee represent.

0%   25%   50%   75%   100%

Source: Survey in Belgium 1988:
Managers and Employee Representatives each N = 587 (weighted), N = 348 (unweighted)

Figure 49: Future employee participation in planning and implementing new technology

Belgium

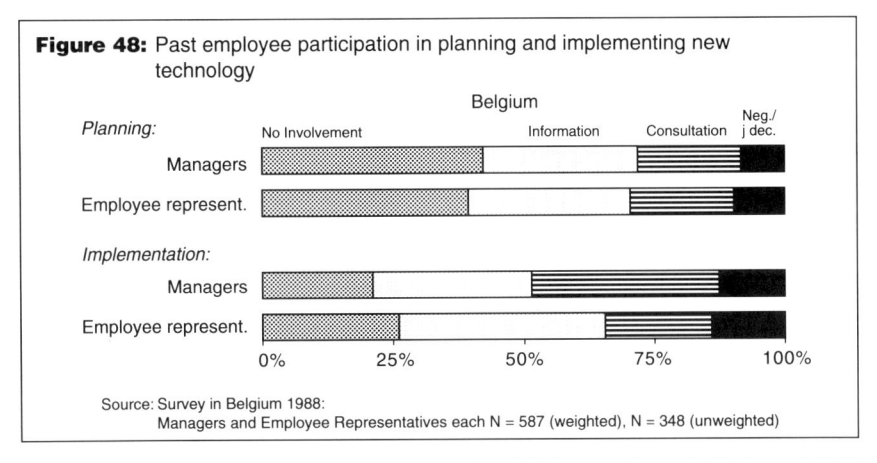

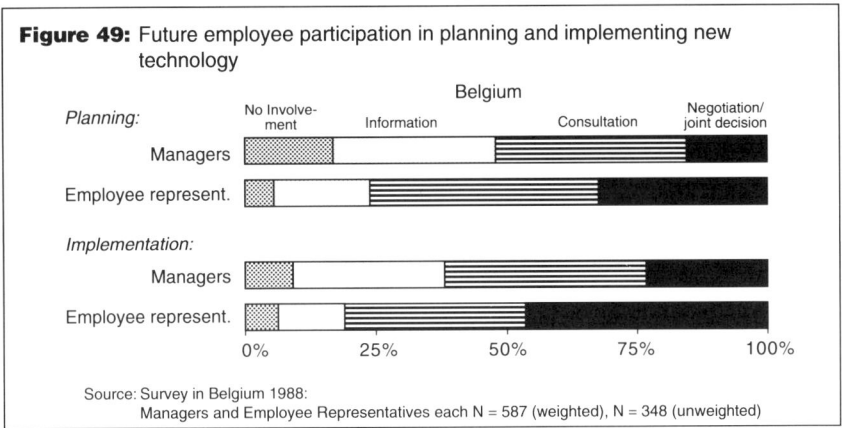

Planning: No Involvement | Information | Consultation | Negotiation/joint decision
- Managers
- Employee represent.

Implementation:
- Managers
- Employee represent.

0%   25%   50%   75%   100%

Source: Survey in Belgium 1988:
Managers and Employee Representatives each N = 587 (weighted), N = 348 (unweighted)

There is a high level of unionisation, which, allowing for non-active members (retired, unemployed, students etc.) is around 75%. Moreover, the Belgian unions have a strong presence at company level; almost every medium-sized company (of at least 100 employees) has a company council (dealing with social matters) and an occupational health and safety committee. In many cases trade union officials are members of both committees and are often regarded as the sole representatives of employees. Unions and union negotiators can therefore rely on a strong legitimisation of their position in talks with employers or their representatives.

There are two conditions about the Belgian industrial relations system which create a favourable setting for a national technology agreement. Belgian agreements are concluded at national level, without the direct participation of the rank and file; there are also agreements at enterprise level which supplement those concluded at national level. The consensus about the content and form of industrial relations, with the emphasis on bargaining at national level and with high union membership means that unions are strong enough at plant level to push through such agreements.

However, whilst these two conditions are conducive to creating a favourable setting for a national technology agreement, the conclusion of the 1983 technology agreement in Belgium has to be seen in the light of an increasing interventionist role of the government in industrial relations in the early 1980s. Both the form and content of bargaining were subject to government intervention, particularly in relation to incomes policy, deregulation, modifications to index-linking, imposition of flexitime etc. Indeed, the Belgian government played an important role in defining the form of the 1983 agreement by introducing textual suggestions and imposing a deadline for conclusion of the agreement.

The main provision of the National Technology Agreement is that employers have to initiate procedures to inform workers on economic, financial or technical factors that are associated with the introduction of new technologies, the nature of its social consequences and its timing. In principle, this information has to be in writing, but it is mandatory for social factors, and the procedure should start at least three months before the actual introduction.

As a result of government intervention in the drawing up of the agreement, the text has a lot of intrinsic difficulties, problems with definitions and concepts, along with several procedural problems. This has meant that the agreement has had a very limited role in ensuring that adequate information on new technology is given to employee representatives in both the planning and implementation stages of technological change.

Having outlined the context of industrial relations and the existence of the national technology agreement in Belgium, how do our results compare with our expectations?

If we consider the planning phase of technological change there is a striking similarity between in the responses given by management and employee representatives in the survey. 42% of managers and 40% of employee representatives agreed that there was no involvement in this phase of new technology; the percentages for information were 30% and 31% respectively. 19% of managers and exactly the same percentage of employee representatives said that the planning phase was characterised by consultation.

There was little difference in opinion in the responses of both parties on the implementation stage of technological change. Here the percentages were no involvement (managers 21%, employee representatives 26%), information (managers 32%, employee representatives 39%), consultation (managers 33%, employee representatives 21%).

In comparison with all the other EC member states, Belgium is ranked in the upper half in the degree of employee participation in both the planning and implementation phases. If we assess the Belgian results in the light of the Val Duchess Joint Opinion — which recommends that employee representatives should at least be consulted about technological change — just over a quarter of all managers state that participation in the planning phase of technological change consists of levels of consultation or more intense forms. Just under a half of all managers claimed that similar participation levels occurred in the implementation phase. These are low figures given the existence of the National Technology Agreement in Belgium.

If we examine the role of works councils in Belgium, we can see why participation levels in technological change are not higher than they are. Works councils were established by a law of 1948, and this law has undergone revisions since then either by national agreements or by Royal Decrees. The most important modification was the 1973 law concerning the right of information. This applies to all private sector enterprises and was extended in 1975 to cover non profit-making bodies and similar organisations. A works council must be established from the moment a firm has 150 employees. It is a joint body, chaired by the head of the enterprise or his or her representative.

Employee representatives vary according to the size of the enterprise but the legislation on works councils in Belgium provides for no specific role for trade unions as such. However, despite the exclusion of trade unions from the workplace, trade unions regularly submit lists of candidates for the four-yearly elections. Works councils have three main tasks: consultation, co-determination and the communication of information. Works councils have the right to be consulted on all areas affecting work organisation and conditions, vocational training and productivity, projected redundancies and they can examine redundancy provisions and the criteria for selection for redundancy. Co-determination only takes place on hours of work, job evaluation, holidays, and welfare. Works rules are normally agreed on unanimously by the Council. The 1973 Royal Decree required the regular disclosure to councils of information concerning the economic and financial position of the undertaking. In sum, Belgian works councils are essentially advisory only and therefore their influence in promoting strong forms of participation in technological change is limited.

How can the results for Belgium be explained in the light of the Belgian national new technology agreement? In a major study of the Belgian national new technology agreement (CA039)[13] the authors concluded that despite the widespread knowledge of CA039 by shop stewards, they were able to obtain more information on new technology by using other legal instruments and laws than by the use of CA039 alone, regardless of the subject matter concerned. This was true on matters such as employment, occupational health and safety, skills, work organisation, wages, economic matters and technical information. Secondly, once future employment had been safeguarded, only minor interest was directed towards other topics. Finally, and most importantly, most of the important managerial decisions had already been made before the three month consultation procedure laid down by the agreement had commenced. The authors concluded that: "...on the whole, the Belgian technology agreement was little used and did not solve the issue of workers' participation or union influence over new technologies".

What conclusions can be made about the Belgian survey results which show that, despite the existence of a national technology agreement, Belgium does not feature as prominently as one might expect among those EC member states with a high degree of participation? The obvious conclusion that can be made is that such a national agreement provides no guarantee of effective participation. Plant agreements on new technology — even allowing for the fact that many of the issues thrown up by technological change in the workplace do not always lend themselves to the agenda of a formal agreement — are often more effective.

There are few plant agreements on new technology in Belgium, and most of these were negotiated before the emergence of the national agreement on new technology. In any case, they mostly cover the same subjects and their procedures are very similar. Given their rather marginal use, some of these agreements are little more than a formalisation of bargaining practices which already existed anyway. The extent of employee participation in Belgium in our survey is thus a reflection of a combination of legislation and the bargaining power of trade unions, both of which existed before the conclusion of the national technology agreement.

What forms of participation do managers and employee representatives envisage for the future? The first point to note is that there is a substantial shift in opinion by both sides in favour of higher forms of participation. In the planning phase of technological change just over half of all managers are prepared to carry out the Val Duchess Joint Opinion guidelines.

In the implementation phase three out of five managers are prepared to do so.

Employee representatives have higher expectations; in the planning phase three out of four of them want Val Duchess Joint Opinion guidelines to be observed, but 42% of our total number of respondents thought that participation should be consultative. In the implementation phase, the percentage subscribing to Val Duchess forms of participation was higher

— 82% — but again, a large number of these employee representatives favoured consultation. Only a third of Belgian employee representatives subscribe to ETUC guidelines on resolving technological change by negotiation or joint decision-making.

There is some degree of disagreement between the two sides on the forms of participation that they desire for the future, although the differences between the two sides are not great.

In comparative terms, Belgium ranks in the upper half of the tables both in the planning and implementation stages of new technology introduction. Its position in respect of past and future levels of participation is very similar to that of the Netherlands. We can summarise the Belgian case by using our explanatory framework in Table 8.5 below:

TABLE 8.5

Variable

Technological Objectives	Fairly high dependence by management on the skills and problem-solving abilities of employees
Management Style	Generally co-operative; well-established works council system but its role in enabling employees to influence technical change is limited; National Technology Agreement not always used and is not very effective
Bargaining Power	Trade union density high; links with works councils strong; fairly technically knowledgable and skilled membership; language divisions overlay inter-union rivalry between the CSC (Catholic) and the FTGB (Socialist)
Regulation	Legislation which provides for works councils; collective agreements provide the main means of dealing with the implications of technological change
Industrial Relations System	Fairly centralised but Belgian employers increasingly fostering more decentralised bargaining; low strike incidence; extensive government powers to intervene in collective bargaining

6. U.K.

The U.K. is alone in the European Community in not having a formal institutionalised system of industrial democracy at company level in the private or public sector. Generally speaking, British trade unions have been interested only in participation schemes which are clearly integrated into the established union structure and envisage some sort of parity representation, whereas management is strongly opposed to any union influence. However, the shop steward movement has secured extensive informal participation at the level of the workplace, but efforts during the 1960s and 1970s to create more formal arrangements came to nothing.

In our survey, compared to other European Community countries, Britain has a degree of participation around the middle of our table in both the planning and implementation stages of new technology. In the planning

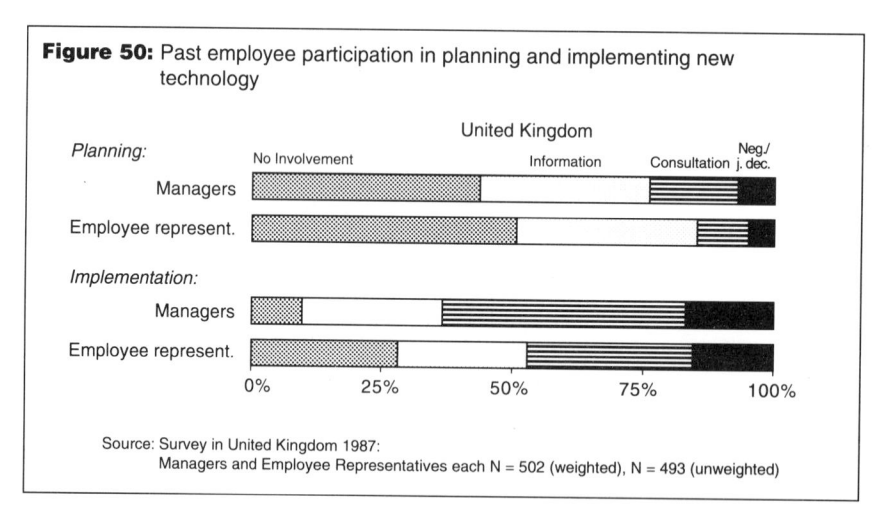

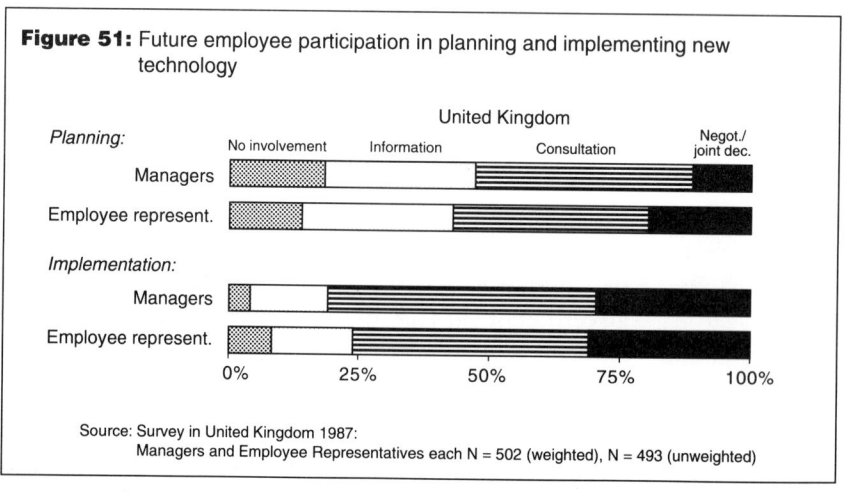

Figure 50: Past employee participation in planning and implementing new technology

Source: Survey in United Kingdom 1987:
Managers and Employee Representatives each N = 502 (weighted), N = 493 (unweighted)

Figure 51: Future employee participation in planning and implementing new technology

Source: Survey in United Kingdom 1987:
Managers and Employee Representatives each N = 502 (weighted), N = 493 (unweighted)

stage 7% of managers report that negotiation/joint decision-making takes place in new technology introduction and 17% of managers report that it takes place in the implementation stage. Employee representatives percentages for negotiation/joint decision-making are similar (4% in the planning stage and 15% in the implementation stage). When it comes to information/consultation however, there are similar responses for the planning stage but there is some difference of opinion in regard to the implementation stage; 49% of managers report that information/consultation takes place in the planning stage (employee representatives 45%) but in the implementation stage 73% of managers state that information/consultation takes place whereas only 57% of employee representatives agree.

How do the British results compare with the guidelines which emanated from the Val Duchesse joint opinion? Just under a quarter of British managers report that employee representatives are consulted or higher levels of participation exist in the planning stage of technological change, and in the implementation stage this figure rises to 63%. What is particularly notable about this 63% figure is that it is largely made up of consultation. This indicates that in the United Kingdom managers there have a high dependence on the co-operation of the workforce in operating new technology. It also reflects the growing importance in the U.K. of joint consultative committees which have grown in importance in the 1980s (see below).

How can these levels of participation be explained in the light of our explanatory framework which was outlined at the beginning of this chapter? If we consider the five main factors which shape the opportunities for employee involvement in new technology introduction, it can be argued that four of them are unfavourable, and only the dependence of management on workforce co-operation in the implementation stage of technological introduction can be said to be favourable.

British management lags behind its counterparts in both France and Germany in its technological innovation in terms of the re-organisation of its production equipment, and the replacement of old special purpose machinery with new automated equipment of a highly complex type. There are several reasons which can be cited for this; a lack of engineering expertise among management generally; a lesser availability of investment funds and the short-term orientation of companies' financial organisation and accounting philosophies which induces managers to pursue quick financial returns from technological investment.[14]

At the same time, although the comparatively high level of unemployment in Britain makes available a large pool of labour, this labour is not generally of the right type to restructure work organisation in line with the new production concepts. This is largely due to the decline of the apprenticeship system which has meant that skilled labour is in short supply. Moreover, the traditional skilled labour that is available is not always of a type which can be flexibly deployed. In such circumstances management in Britain have had to both invest resources in initial and further retraining and to overcome union resistance to the elimination between crafts. Thus there is evidence of a strong dependence on the part of management on the co-operation of its workforce in the implementation stage of technological introduction.[15] However, this means only a very loose form of informal participation by the workforce and it is often carried out as much to circumvent the influence of trade union representatives as it is to further genuine participation by employees.

During the 1970s there were limited experiments in Britain with formal participatory schemes in the steel industry and in the Post Office, but for various reasons these did not last for long. In 1977 the Bullock Committee on Industrial Democracy which was set up by the Labour government constituted an important effort to achieve a more general form of formal co-determination at company level. However, neither the Majority Report, representing the union view, nor the Minority Report, representing the employers' view were implemented. At present, any idea of worker directors does not feature on the political agenda in Britain. British managers are firmly against any participation at Board level and are also opposed to greater participation by shop stewards at a lower level. There is evidence to suggest that shop stewards' committees sometimes function in similar ways to the German works council[16] but that, because management neglects to initiate early dissemination of information and consultation, they are unable to elicit the same involvement in the problems of change as achieved in Germany. Even if management moved towards involvement in decision-making on a wider range of issues, local shop stewards' committees would not have the resources and capacities to respond effectively to such a challenge, nor would they achieve much support in this respect from the external trade union movement.

In recent years there has been a resurgence in joint consultative committees in Britain. Such bodies, as their name indicates, are only devoted to consultation and information and they give workers no co-determination or veto rights. Moreover, such bodies are voluntary and they are only found in a minority of British workplaces — 37 per cent in the early 1980s. Despite their limitations, they are valued by both management and employees as a useful forum of discussion and this fact alone might explain our survey results which show a substantial degree of "information/consultation" in both the planning and implementation stages of technological innovation. The divergence between the views of managers and employee representatives in the implementation stage might be accounted for by management's greater faith in "information/consultation" as an adequate means of introducing new technology than employee representatives who are more likely to be in favour of negotiation or joint decision-making.

Prior to the beginning of the 1980s shop stewards in Britain — particularly craftworkers — had traditionally built up a great deal of influence over various aspects of their work environment (allocation of work tasks, pace of work, manning, recruitment of apprentices, grading, piece-rates and the deployment of labour). Much of this influence was based on informal understandings between workers and management, established by long-standing custom. British workers' representatives have rarely sought, and have not been given, any participation rights as far as major economic change is concerned, which invariably includes new technology. The lack of such rights, coupled with the absence of any legal safeguards made workers particularly vulnerable to the forces of change that they had to face during the 1980s. As a result, their influence has been considerably eroded by a combination of anti-union legislation, major changes in the labour market, the contraction of the manufacturing sector, the erosion of full-time male, manual employment, the growth of the service sector and the creation of a large number of part-time jobs. In such circumstances, loose customary rights which are based on informal understandings have been much more vulnerable to unilateral management withdrawal than those which are guaranteed by law. Thus, although British workers enjoyed a degree of influence over their work environment during the period prior to the 1980s which in many ways was comparable to, and sometimes better than their German and French counterparts, the balance has swung decisively in favour of management.

In sum, whilst there is some evidence that management has to rely on its workforce to achieve its objectives for introducing the new technology, the marked absence of any legislation in Britain to provide for institutional arrangements for participation generally has meant that trade union representatives and their members have been particularly vulnerable to the adverse economic and political changes that have taken place during the 1980s. The organisation of the British trade union movement itself, with its complex structure of overlapping and inter-linking forms of organisation cutting across occupations, skills and industrial boundaries does not lend itself readily to coping with microelectronic technology. The standard pattern of multi-unionism at plant level means that it is difficult to achieve a common approach to a technology which blurs occupational and skill boundaries at the work-place.

The response of the British trade union movement to the new technology has been characterised as adaptive and retrospective — very much in line with past responses to earlier technological innovation. The raison d'être of British trade unions has always been primarily concerned with collective bargaining both as a means of maintaining and improving the terms and conditions of employment of their members and of limiting managerial prerogative. The approach has been largely defensive, concentrating on job losses, pay and working conditions. Thus it is not surprising that British unions believed that new technology issues could be subsumed under the aegis of collective bargaining by the negotiation of new technology agreements.

The new technology agreements that were signed in Britain were modest, and largely limited to the white-collar sector. There seems to have been a great deal of resistance by employers, both individually and collectively, to such agreements and such agreements are few and far between. It appears that the recession, the erosion of large parts of manufacturing industry which had traditionally provided the 'heartlands' of shop steward organisation, and a government committed to reductions in public spending and set against any kind of planning for industrial and economic expansion have all combined to severely limit the ability of trade unions to influence the direction and form of the technology by means of new technology agreements (NTAs). Indeed, NTAs are now less common than they once were in the early years of the 1980s period, and where they do exist some of them are seen by management as facilitative devices which give them a carte blanche to introduce wide-ranging technological changes in the future. In addition, trade union density in Britain has fallen from around 51% in 1979 to around 42% in 1989.

What degree of participation in technological change do British managers and employee representatives prefer for the future? If we take the planning phase of new technology and consider the views of British managers, we can see that there is shift in the preferences of managers for a greater level of participation. The percentage of managers who reported "no involvement" of employee representatives in the past (44%) declines to 18% for the future; whereas 23% of managers stated that participation was characterised by consultation or higher levels in the past, 53% would prefer these forms of participation in the future. Again it is notable that this 53% figure is largely made up of consultation, and it illustrates a strong preference on the part of British managers for simply consulting employees rather than entertaining any stronger forms of participation. Indeed, the United Kingdom ranks 10th — above Italy and Portugal —

in managerial preferences for future participation in the planning stage of technological change. So, while the U.K. falls within the general European average for satisfying Val Duchesse guidelines in the future, our data reflects the determination of British managers to maintain their prerogative in deciding on major decisions on new technology.

In the implementation stage British managers show a slight improvement in the level of participation that they desire for the future. However, one must qualify this by noting that the shift in their attitudes is largely one of no involvement whatsoever to either the provision of information or consultation (see Figures 50 and 51).

A similar picture can be shown for employee representatives in Britain: 51% reported "no involvement" in the planning stage of new technology in the past but, not surprisingly, this percentage declines to 14% for the future. There is little disagreement between both sides about their preferences for the future. What is significant is that British employee representatives lag well behind their employee representative counterparts in the rest of the European Community in seeking negotiation or joint decision-making as a means of influencing technical change; in the planning stage they are ranked 11th (above Portugal) and in the implementation stage they are ranked 9th. Not only are they well below their European counterparts in supporting ETUC policies, but only a small number of them follow the guidelines of Val Duchesse.

The British situation can be summarised by examining the influence of our five explanatory factors in Table 8.6 below:

TABLE 8.6

Variable

Technological	Some dependence by management on the skills and problem-solving abilities of employees in the implementation stage; poor training provisions; emphasis on increasing flexibility of labour; "short-termism"; shortage of skilled labour
Management Style	Generally paternalistic; shift from negotiation to consultation in the 1980s; emphasis on moving from collectivism to cultivating "individualism" by persuading employees to support management policies; hostility to formal institutionalised forms of participation; reluctance to negotiate new technology agreements with trade unions; management vigorously opposed to further improvement in levels of participation through legislation

Bargaining Power	Eroded substantially during the 1980s by labour market shifts and anti-union policies of government; inter-union rivalry; multi-unionism; little cohesion between different unions towards technological change; trade union emphasis on pay and job security; little concern for quality of working life; shop steward influence at workplace level still important; little emphasis on training employee representatives in new technology; shop stewards now isolated from wider trade union movement and external labour market
Regulation	No legislation whatsoever providing for formal institutionalised systems of industrial democracy at company level; trade union preference for subsuming technological change under collective bargaining; informal influence on technological change by shop stewards
Industrial Relations System	Decentralised; heavy reliance on collective bargaining; still traces of "voluntarism"; high level of industrial conflict

7. FRANCE

There is perhaps more fragmentation among the trade unions in France than in any other European Community country. The ideological divisions in the international labour movement are reflected and exaggerated in France more so than in any of the other EC countries. This is one reason why trade union density is low in France (15-18%). However, French trade union density is difficult to measure; there is a large discrepancy between the membership claimed by the unions themselves and evaluations made by employers' associations. There are five major trade union confederations which are considered as being representative at national level: the CGT (Confédération général du travail); CFDT (Confédération française démocratique du travail); the FO (Force ouvrière); the CFTC (Confédération française des travailleurs chrétiens) and CFE-CGC (Confédération Française de l'Encadrement). In general, there is little unity between these confederations.

French unions have rarely built up bureaucratic organisations compared to their counterparts in Germany, for instance. In the CGT in particular, there has traditionally been more emphasis on having an active core of 'militant' organisers, rather than recruiting a stable mass membership. These militants in the past have preferred to foster strikes and political

Figure 52: Past employee participation in planning and implementing new technology

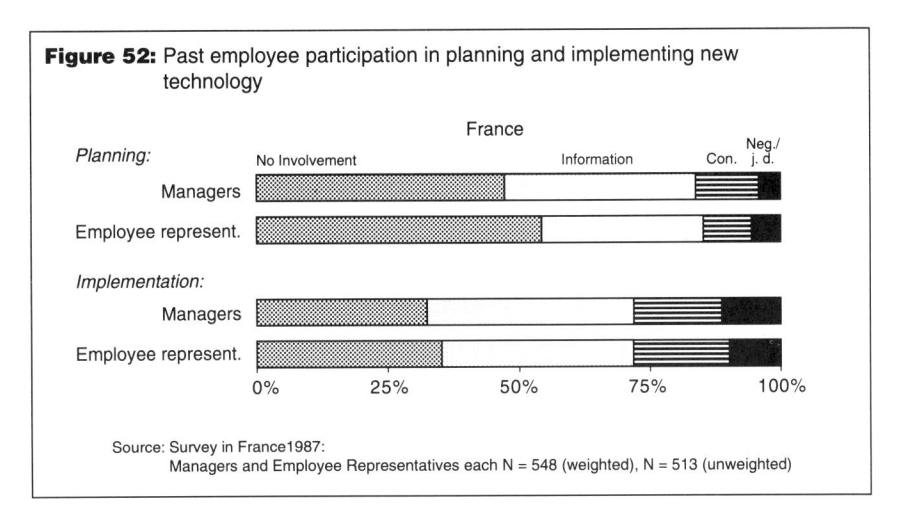

France

Planning: No Involvement | Information | Con. | Neg./j. d.
- Managers
- Employee represent.

Implementation:
- Managers
- Employee represent.

0% — 25% — 50% — 75% — 100%

Source: Survey in France 1987:
Managers and Employee Representatives each N = 548 (weighted), N = 513 (unweighted)

Figure 53: Future employee participation in planning and implementing new technology

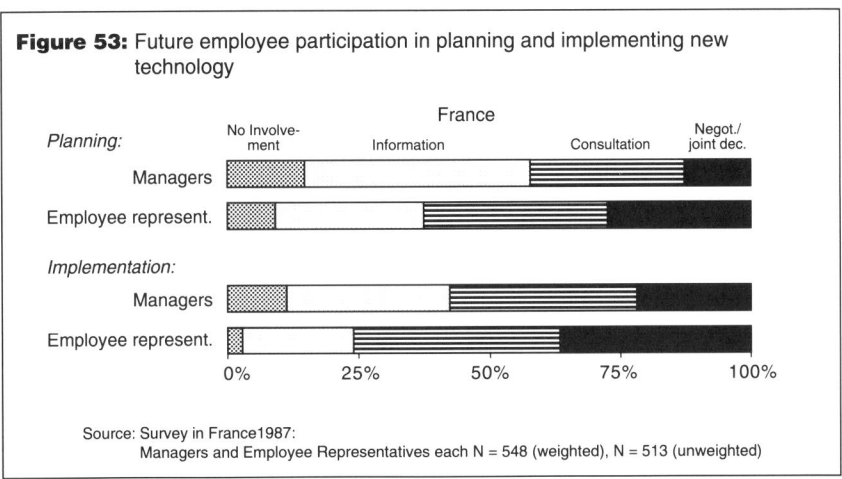

France

Planning: No Involvement | Information | Consultation | Negot./joint dec.
- Managers
- Employee represent.

Implementation:
- Managers
- Employee represent.

0% — 25% — 50% — 75% — 100%

Source: Survey in France 1987:
Managers and Employee Representatives each N = 548 (weighted), N = 513 (unweighted)

action rather than engaging in collective bargaining with employers, which they saw as class collaboration. Not only is there a great deal of inter-union competition, but also collective bargaining has not been the main method of rule-making — although this is gradually changing as a result of the Auroux reforms.

In contrast to the plurality of unionism, at national level the employers are more united — in the CNPF (Conseil national du patronat français) — than the various trade union confederations, and embraces three quarters of all French enterprises. The CNPF does engage in collective bargaining, though not on wages, which are regulated at industry-wide level. Minimum salary levels are set by the French government. The CNPF pursues an active social policy; it seeks to improve training and the development of employees and believes in communicating directly to employees and not through unions. The employers also aim to improve the quality of working life through various means, including quality circles and the individualisation of the employment relationship, in order to reduce the number of issues that are confronted by collective action.

State intervention in industrial relations in France is substantial. This reflects the traditional reluctance of employers to use voluntary collective bargaining, although the unions themselves vary in their attitude to collective bargaining as a means of job regulation. The Auroux report in 1982 led to legislation which sought to promote more collective bargaining and to engender a climate whereby employers would become more aware of their social responsibilities and unions would become more attentive to economic constraints.

Given the industrial relations system which has just been outlined above and the factors which we identified as shaping the extent of employee involvement in technological change at the beginning of this chapter, we would expect France to have a relatively low degree of employee participation in decision-making in technological change. The French system of industrial relations is characterised by a management style which is often described as a mixture of autocratic and paternalistic elements. Management eschew any participation of workers in decision-making and insist on managerial prerogative on nearly all important issues. There is an unwillingness to involve subordinates in decision-making in any meaningful sense and a low level of concern for gaining the consent of those affected by decision-making. This authoritarian management style is reinforced by the ideology and policy of some of the major union confederations, epitomised by the open rejection by these unions of management goals. In addition, the weakness of the unions in both numerical and organisational terms does little to convince management that it needs to modify its adversarial style. Although there have been recent moves towards more participative forms of work organisation as a result of the Auroux reforms of 1982, the transition to a new model of industrial relations in France is proving difficult to achieve.

Thus there are at least three of the factors which we identified earlier which are unfavourable for worker participation in new technology in existence in France — management style, union power and regulation. These factors are reflected in our survey results which show a close similarity between the responses of both managers and employee representatives.

In the planning stage of technological change 47% of managers and 54% of employee representatives stated that there was no involvement. This represents a high level of no involvement by French employee representatives; in other countries of the European Community only

Spain, Greece and Portugal have higher levels. 37% of managers and 31% of employee representatives specified that participation was concerned with the communication of information. Only 15% of managers stated that participation involved levels of consultation and above, in line with Val Duchesse Joint Opinion guidelines.

Similar low levels of employee participation were found on the implementation stage: "no involvement" (managers 32%; employee representatives 35%); "information" (managers 39% employee representatives 37%). Val Duchesse guidelines of consultation and higher levels were reported by 29% of managers.

The low degree of employee participation in France can be partly accounted for by a brief examination of the institutions that were created as a result of the Auroux laws which were enacted by the Socialist government in 1982. Although these reforms strengthened workers' formal participation rights in several significant respects to what had existed prior to the election of Mitterrand's Socialist government, actual participation so far remains unimpressive by European standards.

Unlike Germany and Denmark, there is no parity co-determination at company level in any economic sector. In private limited liability companies only two members of the works committee (Comité d'Entreprise) may attend the supervisory board in an advisory capacity; in the nationalised industries there is a tri-partite board, with both government and worker representatives having a third of the seats and exercising the same rights as other directors. Participation at plant level through the Comité d'Entreprise is also limited. This is due to the fact that the law grants workers' representatives only few co-decision rights (mainly on social issues) and that rights to consultation and information are frequently ignored by management or interpreted in a one-sided way. There is a requirement on French employers to inform the Comité d'Entreprise of impending major changes in the enterprise, which includes any important technological change which could affect employment, skill requirements, earnings, training or working conditions. In establishments with over 300 employees the Comité d'Entreprise may be advised by an outside expert.

Whilst in theory there is considerable scope for participation, since the employer must provide adequate documentation one month in advance and must justify any refusal to follow the advice of the Comité d'Entreprise, in practice there are a number of restrictions. There are problems of interpretation. For example, what constitutes "new" technology? What is meant by an "important" change? Several judicial rulings have ruled in favour of the employer on many of these questions. The Auroux reforms added little to the actual power of the Comité d'Entreprises, although most of those elected to such bodies are trade union representatives; decision-making remains in the hands of employers, who are only obliged to consult and not to negotiate with the Comité d'Entreprises.

Another aspect of the Auroux reforms for participation stems from a more innovative provision, namely the right of expression for groups of shopfloor workers (Groupes d'Expression Directe). These aim to allow direct and collective expression by all workers (but not through unions) on "the content and organisation of their work and the implementation of measures conducive to better working conditions in the company". These groups are established, when possible, through agreements between employers and workers' representatives (trade unions or workers' delegates); or, failing that, through the decision of employers, at workplace or company level.

What effect have such Groupes d'Expression Directe had on participation in French enterprises?[17] On the one hand such groups are seen as having a very limited potential for industrial democracy either because of worker apathy or because they are not seen as important by unions in some industries. On the other hand, they at least provide an opportunity for workers who do not question the power of employer decision-making to speak out on matters which affect them in the workplace. A French Ministry of Employment Report of 1986 noted that there had been few volunteers to assume the leadership of the Groupes it surveyed, and sessions were poorly planned; they also found that while the initial establishment of such bodies led to much enthusiasm on the part of both manual and white-collar workers, such enthusiasm rapidly evaporated when the meeting brought no concrete results. As one speaker put it at a conference on such groups: "plus informatrices que transformatrices". The same report concluded that the scope of matters covered by the expression groups were far too narrow in scope and often simply reproduced no more than the wording of the law.

The setting up of these Groupes d'Expression Directe made lower and middle management in French enterprises less authoritarian and more responsive to the views of the workforce, although it is argued that this occurred where such change in management attitudes was taking place anyway.

What role do the Groupes d'Expression Directe play when new technology is introduced into French firms? It is important to stress that while such bodies do sometimes enable workers to develop a new capacity to solve technical problems without any intervention by middle or even senior management, many such problems require the assistance of experts outside the workgroup. Given the wide variations from plant to plant in the way the groups operate (who is in charge, how often do they meet, on which topics etc.), and the level at which they are introduced, it is doubtful whether the groups can play more than a marginal role in employee participation in new technology. In any case, the Groupes d'Expression Directe are explicitly authorised to discuss only questions of work organisation. According to the law, they should not be concerned with general problems such as the introduction of new technology; otherwise they might interfere with existing institutions such as the Comité d'Entreprise.

The Groupes d'Expression Directe often include members of middle management, or are organised with their help. They may therefore become channels for downward communication to facilitate management's own strategy for increasing productivity. In many companies, such groups function effectively as quality circles and are unlikely to have a major role in the introduction of new technology.

Recent comparative studies also explain why France has a comparatively low degree of employee participation in new technology in our survey results. Collective bargaining over new technology in France, especially at industry level, is not well developed. Before 1980 there was no negotiation on the subject outside banking and insurance, although unions had developed strategies on new technology from the mid-1970s. At industry level, new technology is subsumed within the broader bargaining agenda. However, in May 1986 two industry agreements on technological change were approved, in the provincial press and banking.

The latter is particularly important, since it followed two years of negotiation and elaborates the legal requirements for consultation with representative bodies at the workplace. It is designed as a framework for progress at enterprise level, defining the main rules to be observed. This agreement specifies the rules and methods of involvement of the Comité d'Entreprise; the agreement provides that the Comité d'Entreprise should be informed of any new technology project before the final decision on implementation: the information should include details of the scheme, its implications for job structure and content, training, and working conditions. It is laid down that the Comité d'Entreprise should not challenge managerial prerogative or the principle of introducing new technology. The application of the agreement is monitored by a joint committee.

The agreement for provincial newspapers is more concerned with job security and preserving pay levels. It prescribes further national negotiations on job classification and shorter working hours. Agreements at enterprise level on new technology in France are few and far between.

Since our survey was carried out a national new technology agreement on the implementation of technological change has been negotiated between the three major employers' organisations — CNPF, GMPE and UPA — and by three union confederations — CFDT, CFE/CGC and CFTC. The CGT and the FO refused to sign this September 1988 agreement. The purpose of the agreement, which was not binding on the signatories, was to establish "a general framework of orientation"[18] within which sector-level negotiations were to be encouraged. As such it was no more than a loose statement of intentions by both sides and it remains to be seen whether it will provide a genuine impetus for negotiations on new technology in France or improve the level of participation by employee representatives in the future.

What form of participation do the parties in French enterprises anticipate for the future? There is a substantial shift in opinion among French managers in favour of higher levels of participation in technological change. Four out of ten of them are prepared to envisage Val Duchesse guidelines of consultation and higher levels for the planning phase of technological change in the future. In the implementation phase, 58% of managers are prepared to envisage similar forms of employee representative involvement.

Employee representatives also show a shift of opinion towards desiring higher levels of participation in the future. 63% of them would like participation levels of consultation and more intense forms — in line with Val Duchesse guidelines — in the planning phase and over three quarters of them would prefer similar levels in the implementation phase. Insofar as adherence to ETUC policies towards participation in the planning stage of technological change is concerned, only 28% of French employee representatives indicate their support — the second lowest percentage in the whole of the European Community. There is also quite a difference of opinion between both sides about the type of participation that should exist in the future, perhaps foreshadowing conflict in the future.

In summary, the relative underdevelopment of employee participation in technological change in France is due to the fact that neither employers nor the bigger trade union confederations favour it. Consequently, it remains doubtful whether government intervention, particularly through the Auroux reforms, will substantially change this state of affairs. Management vehemently objects to the imposition by the state of structures to promote greater worker participation, and the hierarchal pattern has been entrenched for too long a time to be easily changed by legal means. The opposition of the unions to the institutionalisation of industrial democracy is based on the conviction that it integrates unions into the capitalist system and thus weakens working-class opposition without giving a significant level of control. More militant forms of giving control, which do not jeopardise union autonomy, are favoured but cannot be realised due to the weakness of the French unions, which are divided along political and religious lines and have a low level of union density. The price paid by French workers for the ideological intransigence of the larger union confederations is high. They have little control over any of the conditions governing their work situation and remain dependent on a less than benevolent management.

The influence of our explanatory factors on participation levels in France is summarised below in Table 8.7:

TABLE 8.7

Variable	
Technological	Increasing dependence by management on the skills and problem-solving abilities of employees; technologically innovative management but less successful in adapting designs to the needs of production; shortage of highly skilled and of polyvalent labour; notable increase in labour market segmentation; recent state intervention to increase the supply of skilled production workers
Management Style	mixture of autocratic and paternalistic elements; insistence on management prerogative on all important issues; opposed to increased levels of participation
Bargaining Power	Trade unions split along political and religious lines; low union density; low degree of professionalisation; preference for political action rather than collective bargaining; multiple system of plant representation which has developed in a piece-meal fashion; no stable mass membership; limited role of Comité d'Entreprise
Regulation	Regulation of industrial relations largely in the domain of the state through the use of law; although on paper the Auroux reforms created legislation providing for employee rights through Comités d'Entreprise, in practice unilateral employer regulation is dominant; weak links between trade unions and enterprise bodies
Industrial Relations System	Very decentralised; prominent role of the state; unstable industrial relations; recent attempts to promote collective bargaining through Auroux reforms but pace of "modernisation" of French industrial relations system is slow

8. SPAIN

The current Spanish industrial relations system owes its origins to the return to democratic government in 1977, and compared to most other European Community countries, it is extremely complex. The former Francoist trade union structure was dismantled and since then free trade unionism has developed. The most important trade union confederations nationally are now the Workers' Commissions (Confederación Sindical de Comisiones Obreras (CCOO), close to the Communist Party of Spain (PCE), and the socialist UGT (Unión General de Trabajadores), one of the major trade union confederations which is linked to the Spanish Socialist Workers' Party (PSOE).

In 1977 the former system of works councils was abolished by decree, and staff delegates (forming works councils in large enterprises) are now the sole legal representatives of all the workers, regardless of their trade union affiliation, and the unions compete to secure representation on these bodies. Delegates are elected for four year terms. The status of these bodies is regulated by the 1980 Workers' Statute, as amended in 1984. There was an agreement between the UGT and the government in early 1986 which provided for unions to have board-level representation in publicly owned concerns (APEP).

An important recent piece of legislation in Spain was the Law of Trade Union Liberty which was enacted in 1985. This Act guarantees full freedom to employees (apart from certain designated occupations such as the military, judiciary etc.) to join and to form unions; self-employed workers, the unemployed and the retired may join existing unions but may not form separate trade unions of their own. Unions are free to draw up their own rules and are empowered to engage in collective bargaining, organise activities both on and off working premises, strike, and put forward candidates for election as workers' delegates on works councils in Spanish companies.

One feature of the Act proved to be controversial, in that it favoured the two major trade union confederations at the expense of the smaller unions and those unions which were regionally based. This was the part of the Act which provided for a system of representational status (affecting a union's participation in collective bargaining, representation with government etc.) whereby to be considered "most representative" at national level a union must obtain the votes of at least 10 per cent of staff delegates in the country as a whole. This meant that only the UGT and the CCOO were able to qualify as a result of together winning 74.6 per cent of the delegate seats in the 1986 trade union elections. Of the two, although the UGT won 71,327 delegate seats of the total of 177,484, the

CCOO won a dominant position in the great majority of companies with more than 750 employees and in key public sector areas including the railways, Iberia airlines, telephones, gas and energy and the major banks.

Trade unions in Spain are highly centralised in their structure and government, which is only partially offset by the activities of representative bodies within enterprises. Usually, trade union confederations formulate collective bargaining policy, take part in discussions with the authorities, are represented on the management bodies of public social services, manage the most of the financial resources, and are responsible for trade union discipline.

According to a study carried out by the government-sponsored research institute Centro de Investigaciones Sociologicas in 1989[19], 16.5% of Spanish employees belong to a trade union, representing a total membership of 1.3 million, although the total has been declining recently. The Socialist oriented confederation, UGT, has a total membership of about 490,000 and has experienced an 8.2% decline since 1986; the Communist-oriented confederation, CCOO, suffered a 31.2% drop in its membership, which is now estimated to be about 477,000.

The Spanish Confederation of Employers' Organisations (CEOE) was founded in 1977 and held its first congress in 1985. The CEOE has sought to adopt a policy of consensus with the socialist government and the trade unions, recognising the autonomy of both sides in collective bargaining. It is less centralised than the trade union confederations.

Collective bargaining in Spain takes place at all levels, national, sectoral, regional, company and plant level. The collective bargaining and works council system are linked together in a number of important ways. Apart from the many other functions and rights given to works councils by the 1980 Workers' Statute including joint decision-making on a number of practical matters with managers within the enterprise, works councils also have the right to conduct collective bargaining with the employer. These agreements can regulate a wide range of working conditions, and are formalised in writing. It might be thought that the authority of Spanish works councils and work-force delegates to negotiate such enterprise agreements clashes with an equivalent bargaining power granted contractually to those trade unions which are firmly established within the enterprise. In practice, however, this collective bargaining function has in most cases been assumed by the individual representative bodies and not by trade union representatives in the strict sense.

This system of collective representation at enterprise level, together with the important functions assigned to representative bodies within enterprises by the Workers' Statute, does not mean that there are two separate systems to represent workers with, on the one hand, trade unions and occupational associations negotiating agreements at sectoral level, and on the other hand works councils and personnel delegates negotiating with employers at enterprise level. In practice, the two forms of representation have become so closely linked that they might be considered facets of a single system of representation. This is because of the strong presence of the trade unions on works councils and among work-force delegates; another is the major role attributed to the trade unions in the procedure for electing these representatives; a third is the acknowledgement of the right of trade unions to set up their own representative bodies in workplaces — bodies which are connected constitutionally with works councils when the trade unions concerned are well established in the enterprise.

Spanish law recognises the right of trade unions to put forward candidates in the elections held periodically to appoint the members of these individual representative bodies, in competition with any other independent candidates. Success in these elections is vital for the competing trade unions since electoral success means that they are able to exert influence on the representative bodies and they thus acquire the status of the most representative union. Trade unions have secured 90% of all elected posts and there is a continuing trend for independent candidates to be ousted at the expense of trade union candidates. Trade unions are thus firmly established at enterprise level. Another factor which has added to trade union influence at enterprise level was the 1984 amendment to the Workers' Statute, which made it possible for unions to set up trade union branches within enterprises, with broad powers to co-ordinate trade union activity within the workplace. Trade union "delegates" therefore attend meetings of the works councils and thus ensure the link between the external trade union movement and its representative function within enterprises.

The state under the Franco régime used to play a prominent part in regulating industrial relations, but since 1977 its role has gradually declined. However, since the beginning of the 1980s the Spanish government has been playing a major role in what has been called "social co-operation". This involves a form of tri-partism where the government, in consultation with employers' associations and trade union confederations, seeks to shape the content of collective bargaining to remain within limits which are compatible with its broad economic objectives. In exchange for such co-operation with government, both the employers and to a much greater extent, the unions, have been able to exercise considerable influence over the shape of labour legislation and government economic policy-making.

A Comparison of Past and Future Participation in Technological Change

Until 1986 "social co-operation" worked reasonably well. From 1986 onwards the UGT increasingly distanced itself from the continuing austerity programme of the Gonzales socialist government and consequently moved towards greater unity in action with the CCOO. Negotiations over a new social pact ended in deadlock in February 1987 with the UGT demanding a wage increase of 7 per cent and the employers holding firm to a ceiling fixed at the projected inflation rate of 5 per cent. The CCOO, which was not a party to the previous social pact and did not participate in the negotiations over its renewal, entered the 1987 wage round demanding increases of 7-8 per cent. The breakdown of the social pact, the divergence in bargaining positions between the unions and employers and the increased co-ordination of action between the UGT and the CCOO led to an upsurge in industrial conflict which later culminated in a 24-hour general strike on 14 December 1988. The strike, which kept eight million people (or 90% of the workforce) away from work and brought the country to a standstill, was hailed as an historic event by the unions.

The strike nevertheless failed to force the government into an economic policy U-turn, but the Gonzales government only narrowly won the Spanish general election in October 1989. The narrow election victory made the government and the unions realise that they both needed each other and both sides have become reconciled to moderating their negotiating positions. The UGT and the CCOO recently agreed on a joint proposal of demands known as the Propuesta Sindical Prioritaria (PSP) aimed at securing the resolution of grievances on a step-by-step basis with the government. The UGT is gradually loosening its ties with the Socialist party and, for its part, the CCOO is gradually shedding its communist party tutelage.

Having clarified the complexities of the Spanish industrial relations system, it is now time to turn our attention to our survey results for Spain. The first point to note is that Spain has the third highest degree of "no involvement" in the planning stage of technological change among all the European Community countries — with only Portugal and Greece having a higher degree of no involvement; the same percentage (58%) of both managers and employee representatives agree that this is the case. 18% of managers and 14% of employee representatives state that involvement by employee representatives in the planning stage is characterised by either consultation and/or negotiation/joint decision-making.

In the implementation stage there is again a striking similarity between the opinions of both sides. Perhaps the most important point to note is that nine out of 20 managers and employee representatives state that there is no involvement whatsoever in this stage of technological change; this percentage is, incidentally, the highest of all the European Community countries in our survey. Around a third of both managers and employee representatives claim that employee representatives are given information on technological change. The level of involvement by employee representatives from consultation upwards is also low by European standards.

An obvious explanation for the comparatively low involvement by Spanish employee representatives in technological change is the historical legacy of the repression of trade unions under the Franco régime. Whilst the

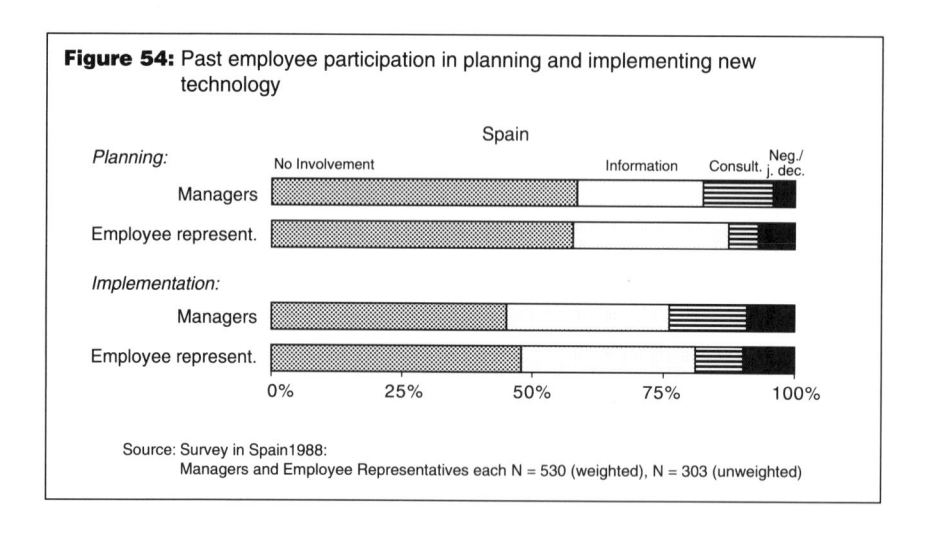

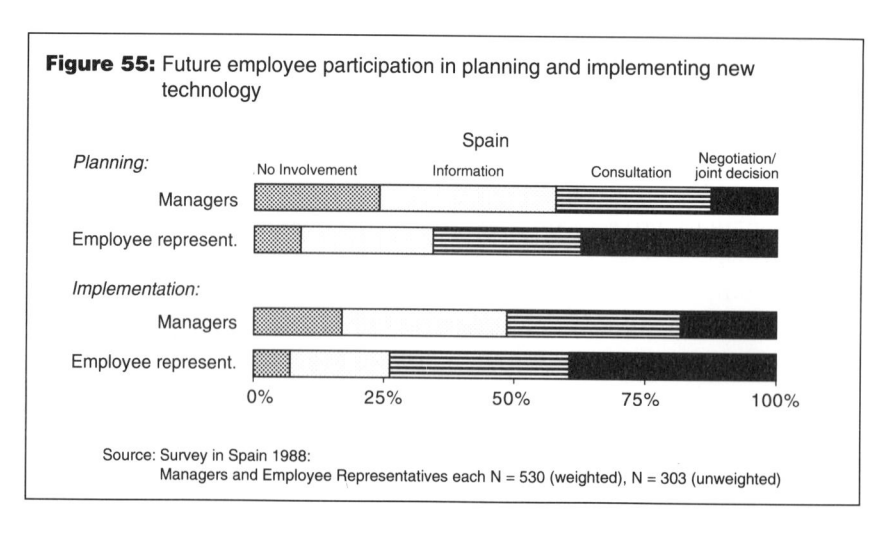

Figure 54: Past employee participation in planning and implementing new technology

Source: Survey in Spain 1988:
Managers and Employee Representatives each N = 530 (weighted), N = 303 (unweighted)

Figure 55: Future employee participation in planning and implementing new technology

Source: Survey in Spain 1988:
Managers and Employee Representatives each N = 530 (weighted), N = 303 (unweighted)

re-building of the Spanish industrial relations system allowed free trade unionism to flourish in the new democratic era after Franco, the trade unions have been heavily reliant on favourable legislation being passed by the government. As we saw earlier, their membership density is still low at 16.5% and after an initial expansion in membership in the late 1970s their growth has stagnated; they are restricted to the traditional working class core and have found it difficult to expand into the growing service sector and to white-collar occupations.

Spain provides a good example of a country where legislation on employee participation provides no firm guarantee that it will emerge in practice. There seems to be a preference on the part of trade unions to pursue their interests through collective bargaining. The employers, however, still guard their managerial prerogative jealously, and whilst they have tolerated the new legislation and have co-operated with the trade unions and the government in economic planning, they are anxious to move towards more flexible and modernised labour practices in order to maintain competitiveness. Spanish trade unions in turn complain that employers have used youth training schemes as a device for hiring cheap, casual, short-term labour.

If we consider the factors which shape the opportunities for employee involvement in technological change which were outlined in our explanatory framework at the beginning of this section, all five factors tend to be on the negative side. There seems to be little need for managers to rely on its workforce to achieve its objectives for introducing technological change; management seems to be less than enthusiastic for participation by employees. In addition, while the Spanish economy has substantially diversified and developed in the last two decades and is more advanced than its neighbouring country Portugal, technological innovation still lags behind that of other major European Community countries.

The trade unions, by their heavy reliance on collective bargaining, have not placed much of a premium on employee participation, and their relations with management are still of a "low trust" nature. There is little evidence that what bargaining power that they have has been particularly effective in influencing technological change; for example, one of the major demands of the joint UGT-CCOO proposal on union priorities is "the extension of employees' rights of information, consultation and participation in company level decision-making".

Finally, whilst there has been a great deal of favourable legislation for trade unions to bargain collectively with employers at all levels since 1977, the legislation has not enabled the works councils to conclude agreements with employers at enterprise level which ensures their involvement in technological change. Indeed, the legislation gives very little in the way of rights for works councils for involvement in technological change; apart from the right to receive reports within 15 days of a managerial decision being announced and to monitor health and safety conditions, their role is very much centred on ensuring that the provisions of the relevant collective agreement are observed.

What degree of involvement in technological change do Spanish managers and employee representatives anticipate in the future? Insofar as the planning stage of technological change is concerned, managers' preferences are much in line with the European average. Those managers favouring "no involvement" declines from 58% in the past to 24% in the future — a significant shift in attitude. There is also a shift in attitude of managers who are prepared to see more information, consultation, and negotiation/joint decision-making in the future. If we consider the percentage of managers who are prepared to offer levels of involvement from consultation upwards in the planning stage, this increases from 18% in the past to 42% in the future, which represents a significant shift in opinion.

There is a great deal of agreement between employee representatives and managers about the level of their involvement in the planning phase of technological change in the past, but employee representatives have greater expectations for the future than their management counterparts. For example, 57% of employee representatives stated that past planning was characterised by "no involvement", but less than one in 10 of them envisage this for the future; the percentage reporting levels of employee involvement from consultation upwards increases from 14% in the past to 66% in the future. 37% of Spanish employee representatives would prefer the planning stage of new technology in the future to be characterised by negotiation or joint decision-making in line with ETUC policies. This places Spain in 4th position in respect to their counterparts in other EC countries. There is a great divergence between the preferences of Spanish managers and employee representatives and one might speculate on the possibility of conflict between the two sides of Spanish industry in the future.

In the implementation phase of technological change, managers report a much higher level of "no involvement" than their European counterparts. 45% of them state that no involvement occurred in the past and three in 10 of them stated that the implementation phase involved the provision of information. However, there is a more liberal attitude evident in terms of their future expectations; the percentage of managers favouring no involvement declines to 17% and half of them envisage levels of involvement by employee representatives of consultation upwards. Employee representatives expect much more than managers in the future; four out of 10 favour negotiation or joint decision-making in line with ETUC policies, and over a third favour consultation.

To conclude, Spain's results for involvement by employee representatives in technological change fall well short of the European Community average in both phases of technological change, and there is a great deal of improvement necessary in order to satisfy the provisions of the ETUC/UNICE Val Duchesse joint opinion. The influence of the explanatory factors in the case of Spain is portrayed in Table 8.8 below:

TABLE 8.8

Variable	
Technological Objectives	Little dependence by management on the skills and problem-solving abilities of employees; management jealously guards its prerogative; management keen to impose modernised working practices in Spanish industry
Management Style	Opposed to increased levels of participation; Spanish industry generally lags behind other European countries;
Bargaining Power	Trade unions still developing following transition to democracy from Francoist period; low union density; tenuous co-operation between UGT and CCOO; centralised union policy; inexperienced trade union officers; heavy reliance on favourable legislation on participation from government; trade unionism concentrated among manual workers and suffering membership decline; tension between unions and government over economic policy; strong presence of trade unions on works councils
Regulation	Legislation providing for works councils as a result of Workers' Statute; extensive links between works councils and external trade union movement; works councils can conduct collective bargaining; trade unions have not placed a premium on participation but rely more on collective bargaining; collective bargaining at national, sectoral and company level; trade unions have board-level representation in publicly-owned concerns (APEP)
Industrial Relations System	Elements of centralisation and decentralisation; conflict-prone; polarisation in Spanish society reflected in "low trust" relations between management and employees

9. GREECE

The Greek industrial relations system, has been largely shaped under the democratically elected governments which ruled Greece following the end of the military junta in 1974. The (conservative) New Democracy government elected in 1974 drew up a new Greek Constitution in 1975, article 23 of which guarantees trade union freedom. Legislation was also introduced by the administration of the Pan-Hellenic Socialist Movement (Pasok) — first elected in 1981 and re-elected in 1985 — which considerably extended trade union rights and freedoms in the private sector in 1982. Further legislation by the Pasok government was enacted in 1983 (covering the nationalisation of certain public undertakings and the right to strike by employees within those undertakings); this was followed by an important Act on Works Councils in 1988.

The right to conduct collective bargaining in Greece is laid down in the 1975 Constitution and the 1982 Law. Collective bargaining takes place at three levels in Greece: national level, covering all workers with the exception of civil servants, agricultural workers and seamen; sectoral agreements at national level (where most bargaining occurs); and local collective agreements at company or establishment level. In the event of failure to agree, the Greek Ministry of Labour can mediate between the parties, and if this process fails, there is provision for the issue to be referred to a Court of Arbitration.

The major form of pay bargaining that takes place in Greece is through the annual general central agreement for the private sector of industry. The agreement is negotiated on the employers' side by the Confederation of Greek Industries (SEV),the General Confederation of Artisans (GSEVE) and the three big cities' Commercial Employers' Associations (ESAPS).

As far as the unions are concerned, the most important body is the Greek General Confederation of Labour (GSEE) which represents around 400,000 employees of the 450,000 unionised workers. Around 30% of Greece's 1.5 million employees are unionised. In effect, GSEE has exclusive bargaining rights for all central level bargaining which almost exclusively determines national minimum pay and hours of work, although some minor groups (such as those in printing and banking) traditionally have had better terms. It is important to point out that there is a multitude of unions in Greece; it was estimated[20] that there are around 3,500 unions in Greece — many comprising as few as 20 or 30 members. The national central agreement establishes the minimum pay of manual and non-manual employees in the private sector in industry and in so doing, also establishes a de facto nation-wide minimum wage rate.

The Greek trade union movement was split as a result of divisions that took place within GSEE in October 1985. The introduction of a package of austerity measures by the Pasok government, including a programme of wage restraint, precipitated a split in the GSEE, which had enjoyed good relations with the Pasok administration from 1981 to mid-1985. The GSEE President endorsed the austerity package; a number of Pasok members of the 45-member GSEE governing council (administrative board), however, announced their opposition to the package and were expelled from Pasok (leading to the expulsion of numerous other trade unionists from the party in the following months).

As a result of this division in GSEE, there were numerous anti-government strikes, of varying degrees of impact, during the late autumn and winter of 1985/86, and dissident unionists on February 9th 1986, set up the Socialist Trade Union Movement of Workers and Employees (SSEK), which supported those members of the governing council who had been expelled. There were subsequent allegations that the 23rd GSEE Congress held in April 1986 had been rigged by pro-government delegates subsidised by the Ministry of Labour. The 24th GSEE Congress in October 1988 was also boycotted by dissident trade union factions. There was even evidence of tension between the pro-government group within GSEE and the government in 1986/87 over incomes policy. There are now a number of dissident trade union centres in Greece and the close links which existed between GSEE and Pasok were not helped by the allegations of corruption that preceded the 1989 Greek general election.

The troubles which have beset the Greek trade union movement do not, however, have a great deal of relevance to workplace issues. The large number of local trade unions in Greek enterprises form the basis of employee representational matters insofar as conditions of work are concerned. The law permits any 10 workers in an enterprise to form a trade union. The 1982 law requires employers, at the request of the union(s), to meet union officials at least once per month to discuss the operational matters affecting the enterprise. In enterprises with more than 80 workers, a meeting room must be provided for union officials, and where an undertaking employs more than 100 employees, a trade union office must be provided. Trade union representatives are elected by the union membership in the undertaking.

Having outlined the basic contours of the industrial relations system in Greece, we now turn to considering our survey results. As we will see later, our results have to be seen within the context of legislation on employee participation, much of which was enacted after our interviewing programme in 1987/88.

Figure 56: Past employee participation in planning and implementing new technology

Greece

Source: Survey in Greece 1988:
Managers and Employee Representatives each N = 243 (weighted), N = 127 (unweighted)

Figure 57: Future employee participation in planning and implementing new technology

Greece

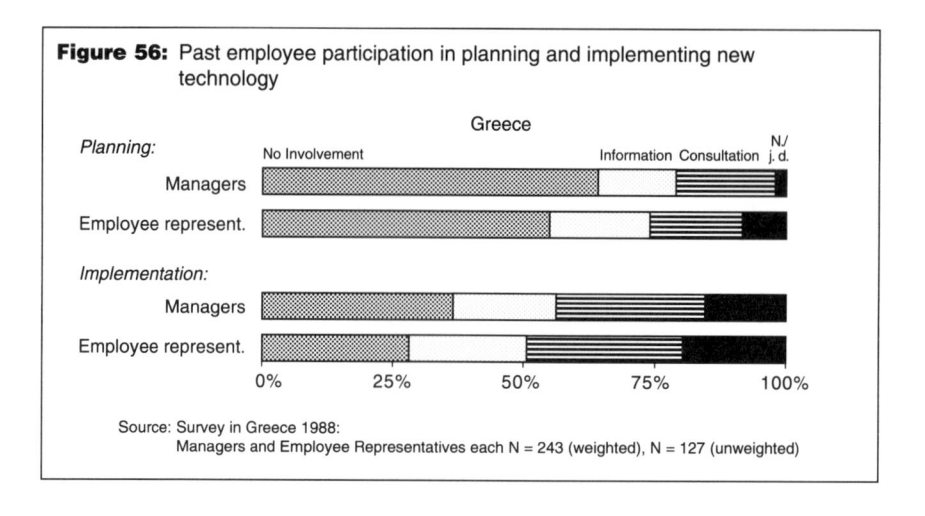

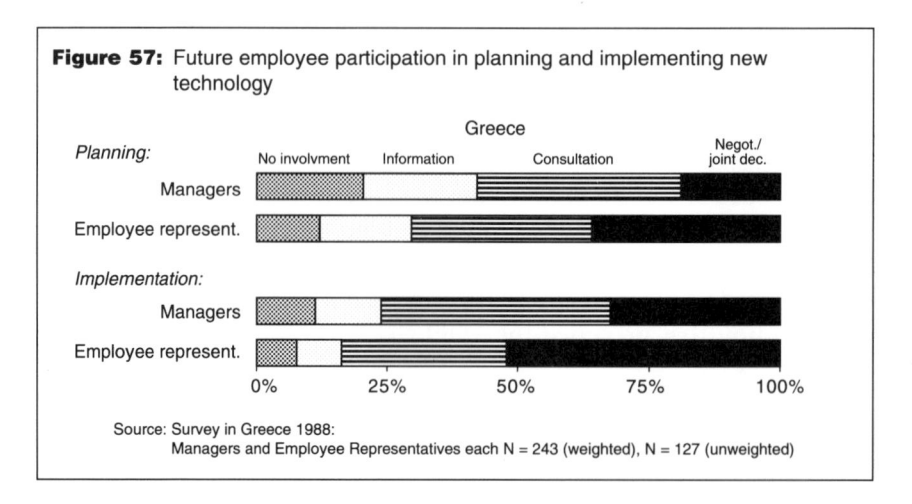

Source: Survey in Greece 1988:
Managers and Employee Representatives each N = 243 (weighted), N = 127 (unweighted)

Our results for Greece show that levels of participation by employee representatives in technological change are very low by European standards. In the planning phase of technological change Greece has the second highest percentage of managers next to Portugal, who reported that there was no involvement at all (65%). Indeed, only one out of five Greek managers stated that employee representative involvement was at a level greater than the provision of information. Employee representatives largely subscribed to this opinion.

In the implementation phase, the results were a little better; here 56% of managers stated that employee representative involvement was restricted to either none or simply to the provision of information — a view which was in line with that of 52% of employee representatives.

How can this very low level of participation be explained in the light of our explanatory framework which was outlined at the beginning of the country-specific section of this report? The most obvious factor to note is that trade unionism in Greece has had little time to develop since its emergence from the dictatorship period of 1967-74 when trade union activity was prohibited by law. Free trade unionism has had little experience in pursuing the concerns of workers with employers and many trade union rights have been artificially promoted by legislation and it takes time for trade unions to adapt to a more favourable environment. Moreover, with the multiplicity of unions, co-ordination and discipline are not easy to attain, while their organisational structure is highly decentralised, making it difficult for GSEE to enter into commitments that would be honoured.

At the same time, the Greek economy manifests the symptoms of unbalanced industrialisation which has for years been associated with significant urbanisation and massive emigration. New technology in Greece lags well behind other countries of the European Community: 95% of establishments employ less than 10 employees — in fact[21] the average number of people employed is 2.3 — and thus the diffusion of new technology is extremely limited in Greek enterprises given the very large number of small firms and the opposition to labour-saving new technology that features in the policies of many unions. The vast majority of such small firms are run on informal lines without any organised pay structures etc.

Moreover, Greece is notorious for its black economy. Greece had the lowest percentage of salaried workers and wage-earners relative to the total active population in the European Community in the 1970s (42%). Once Greek workers acquire a skill and become entitled to higher pay, they leave to become self-employed, which attracts a higher status and is more tax efficient because household income can rise with the use of unpaid family labour. There are almost as many self-employed men in Greece as there are male employees; according to the 1981 Census[22], there are 1.1 million self-employed and 1.2 million male employees.

In the large enterprises in Greece managers have been used to exercising their prerogative for a number of years and might resent legislation on employee participation and trade union rights being forced upon them in a society which has always been polarised in political terms. As we have seen, the Greek trade union movement is very fragmented and its bargaining power within the enterprise is limited. This is not helped by the fact that charismatic leaders with poor training predominate in the unions at all levels.

Until 1988, legislation on employee participation was limited to the Acts which were passed by the Greek Parliament in 1983. One of these Acts introduced a two-tier system in some industries in the private sector and provided for employee representatives to sit alongside shareholders' local authority representatives on supervisory bodies. The role of such bodies, however, was restricted to an advisory capacity.

A second Act covered public sector undertakings, accounting for about 50% of GDP in Greece, including the banks, insurance companies, telecommunications, transport, electricity supply and water. The form that participation takes in each of these industries is laid down by directives from the appropriate government ministry, but generally, there is a supervisory board and a management board in each undertaking. The supervisory board has wide-ranging powers covering investment, finance, project planning and general company strategy. The management board is composed of nine members, three of which are elected employee representatives. There is also a central works council, consisting of nine employee representatives, although its role is as an advisory body only.

Our interviewing programme took place in Greece before legislation was passed on works councils in private sector companies in 1988, and thus we have to discount any effect this may have had on the degree of past participation in our results. However, the impending prospect of such legislation might well have been in the minds of our management and employee representative respondents when they were asked about their future intentions.

What are the essential provisions of the 1988 Works councils law? Legislation on works councils in companies with at least 50 employees (20 if there is no trade union) was passed by the Greek Parliament on 6 April 1988. Under the law, works councils must be set up at the request of employees. The size of the works councils varies according to the size of the company — three members in companies with up to 300 employees, five in companies with between 301 and 1,000 and seven in those with over 1,000 employees. Subsidiary companies are considered as separate entities provided they have 50 employees. If they have less than 50 people, they are represented by the nearest plant with the requisite workforce. The first category is the most common under the law, because as we saw earlier, Greek industry is weighted heavily towards a proliferation of small companies.

The rights of these works councils are fairly wide-ranging, providing for prior information and consultation on company finance, company policy, improvements and changes in working conditions, changes in the legal status of the company, including technological change. Works councils are also consulted on a wide range of matters and also have the opportunity to negotiate with management where both sides are willing to so do.

There is evidence from our survey results that this 1988 works council legislation might have influenced the stated expectations of both managers and employee representatives when they were asked about the level of involvement in technological change that they envisaged for the future. Here the results show a significant shift in opinion on both sides and feature very much in line with the results from other European Community countries covered by our survey.

In the planning phase of technological change, nearly three out of five managers anticipated the level of participation to consist of either consultation or negotiation/ joint decision-making — in line with the ETUC/UNICE Val Duchesse joint opinion. Employee representatives were somewhat more optimistic for the future; seven out of ten of them opted for consultation and higher levels. The views of Greek employee representatives are thus well in line with the mainstream of their European counterparts in supporting ETUC policies.

Similar results were evident in the implementation phase of technological change; here, over three quarters of managers envisaged levels of participation consisting of consultation and above, including a third of managers who favoured negotiation or joint decision-making. Again, employee representatives opted slightly more strongly for higher levels of involvement; 40% of them actually favoured joint decision-making.

In sum, in evaluating the Greek results we have to take into account a number of factors which distinguish Greece from the mainstream of European countries: the very large number of tiny enterprises; the large number of self-employed in the Greek economy; the volatility of its political history; the very limited diffusion of technological innovation; the fact that its trade union movement has only operated in a democratic society for less than two decades and thus its representatives are relatively inexperienced; and the fact that works councils in Greek enterprises are very much in an embryonic state (see Table 8.9 overleaf).

Nevertheless, in one respect at least, our results illustrate a significant shift of opinion on the part of both managers and employee representatives; both parties appear to have found their experience of participation beneficial and envisage levels of involvement in the future very much in line with our results from other European Community countries.

TABLE 8.9

Variable

Technological Objectives	Little dependence by management on the skills and problem-solving abilities of employees; large number of tiny enterprises; sizeable "black" economy; low level of technological diffusion
Management Style	Antipathetic to legislation promoting participation; low level of managerial development
Bargaining Power	Vast multiplicity of small unions; split in GSEE labour movement; strong unionism in banking; inexperienced trade union officers; trade union movement still in embryonic stages after end of military dictatorship in 1974; heavy reliance on favourable legislation on participation from Pasok government
Regulation	Some legislation of a limited form providing for works councils but more extensive legislation passed in 1988; collective bargaining at national, sectoral and company level
Industrial Relations System	Extremely decentralised; conflict-prone; polarisation in Greek society reflected in relations between management and employees

10. ITALY

Italy's industrial relations system has undergone substantial change in the last two decades. During the period from the end of the second world war up to the mid-1960s Italian trade unionism was strongly influenced by a political orientation, with three main union federations in existence. Despite the formation of a unified national trade union centre in June 1944 between socialist, communist and Christian democrat trade unionists as a result of the "Pact of Rome" (when Italy was effectively divided between German and Allied occupation forces) to form the Confederazione Generale Italiana del Lavoro (CGIL), this centre had disintegrated by 1950. From 1950 onwards the present basic tripartite division of the trade union movement developed, with the CGIL dominated by communists, the Confederazione Italiana Sindicat Lavoratori i (CISL) dominated by Christian democrats (but with some socialist influence) and the Unione Italiana del Lavoro (UIL), the smallest of the three, led by social democrats and republicans.

The industrial organisation of the various trade union centres has tended to be weak and under-funded, reflecting their relatively recent development in Italy of "trade" as opposed to "class" unionism and the disunity of the trade union movement; the central organisations have dominated their industrial affiliates. Local chambers of labour, linked directly to their respective trade union centres and uniting workers regardless of industry, have been of more significance than the industrial unions. During the 1960s relations between the main three trade union centres markedly improved and a CGIL-CISL-UIL working alliance was formed in 1972, although the federations retained their separate identities. The high point of both inter-union and national tri-partite unity was reached in 1983. The CGIL-CISL-UIL federation collapsed in 1984 as a result of disagreement over incomes policy, but has since been partly restored.

The model of Italian trade unionism institutionalised in the 1970s (involving a synthesis of political or class unionism with a strong if decentralised base of plant organisation and worker militancy) entered a period of crisis in the 1980s. The worsening economic situation coupled with the emergence of new independent unions of mainly professional employees (sindacati autonomi) in the public sector, the service sector, the health service and the airline industry, led to a decline in the hitherto dominant position of the three main federations. Since 1986 even more localised comitati di base (grass-roots committees), known popularly as "Cobas", have formed in areas such as teaching and essential public services in opposition to not only the main union confederations but the autonomous unions themselves.

As a result of all these problems the three main union confederations lost over a million active members from the high point of nine million in 1980, which amounted to a decline in the unionised proportion of the workforce from nearly 49 per cent in 1977 to just under 40 per cent in 1986. After a substantial loss of membership following the inter-union divisions of 1984 membership began to pick up from 1986 and is now claimed to be near the nine million mark again. An estimated additional four million workers belong to the autonomous unions.

The breakdown of the CGIL-CISL-UIL federation was never total, however. There was a joint signatory agreement on union participation in policy-making in the Institute for Industrial Reconstruction (IRI) in December 1984 and a joint programme of demands for bargaining purposes in July 1985. In December 1985 the three confederations reached a framework agreement with the government concerning all aspects of pay and conditions in the public sector. This agreement

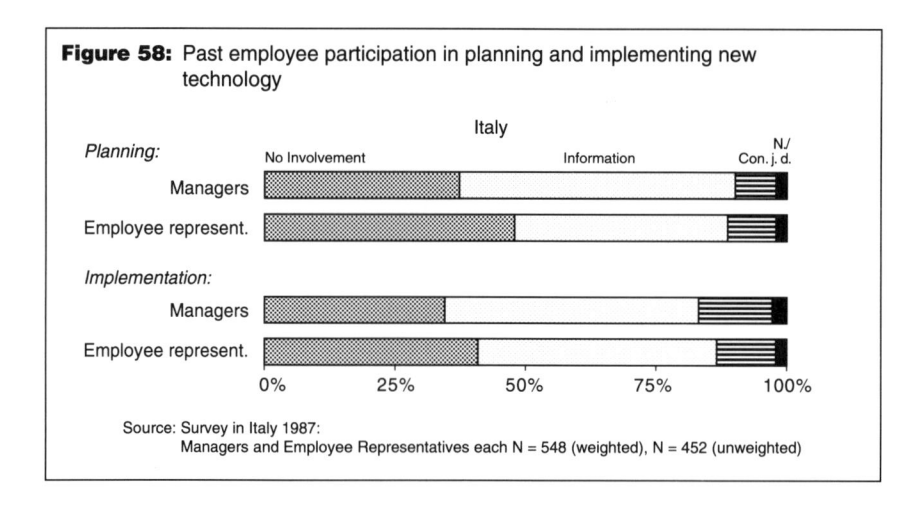

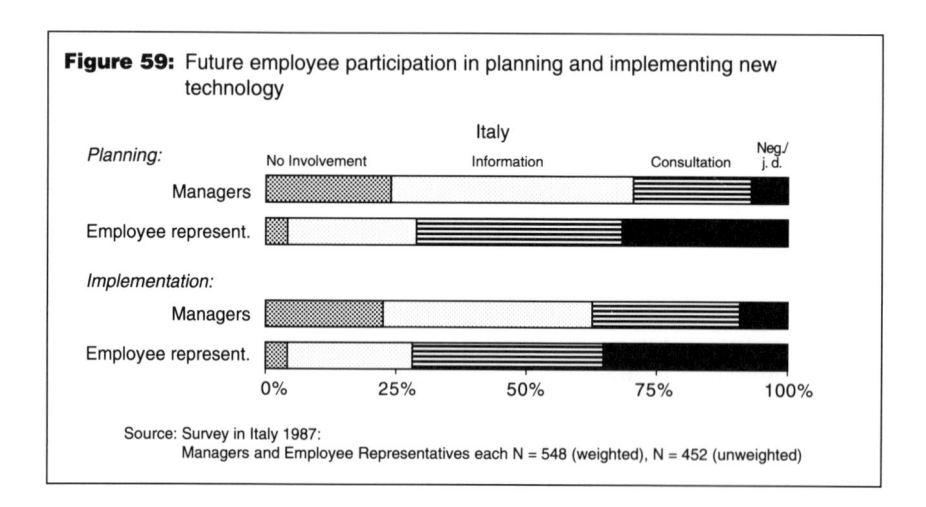

Figure 58: Past employee participation in planning and implementing new technology

Italy

Planning: No Involvement | Information | Con. j. d. | N./

Managers
Employee represent.

Implementation:
Managers
Employee represent.

0% 25% 50% 75% 100%

Source: Survey in Italy 1987:
Managers and Employee Representatives each N = 548 (weighted), N = 452 (unweighted)

Figure 59: Future employee participation in planning and implementing new technology

Italy

Planning: No Involvement | Information | Consultation | Neg./ j. d.

Managers
Employee represent.

Implementation:
Managers
Employee represent.

0% 25% 50% 75% 100%

Source: Survey in Italy 1987:
Managers and Employee Representatives each N = 548 (weighted), N = 452 (unweighted)

included modifications to the scala mobile (the system of wage indexation in Italy), greater workers' participation in policy making, provisions to reduce hours and increase employment, and a voluntary code of conduct for regulating strikes. The agreement was immediately extended unilaterally to the private sector by the employers' confederation (Confindustria).

Collective bargaining in Italy at national level (involving the CGIL, CSIL and UIL, the public and private sector employers' associations and where appropriate the government) occurs irregularly and is concerned with a broad range of issues, notably the system of wage indexation (the scala mobile). Sectoral negotiations, involving industry level unions and employers' associations and typically occurring on a three-yearly basis, set agreements covering all aspects of labour relations and apply to all workers, whether they are unionised or not.

At workplace level, shop stewards' committees — factory councils (consigli di fabbrica) or councils of delegates (consigli dei delegati) — have been established in large numbers since the late 1960s, as part of the development of policies of unity between the CISL, UIL and CGIL, and extend a form of organisation known as commisione interne. However, many of these bodies were not always established through union initiative but were developed by workers in opposition to official union structures, although recently they have been increasingly elected on union "slates". These bodies, which are elected by all the workers, both union and non-union, further define and modify the content of sectoral agreements and they, rather than the unions separately, conduct all intra-plant negotiations. Certain rights to factory-level representation are now enshrined in the Workers' Statute (Statuto dei lavoratori 1970). Bargaining in general is relatively decentralised, and this has allowed for considerable flexibility in adapting to technological change in some sectors.

Our survey evidence indicates that Italy is well below the European Community average for participation in technological change. Nearly four out of every ten managers report that there is "no involvement" by employee representatives in the planning stage in the introduction of new technology and nine out of every ten managers state that employee involvement is limited to no more than information. The situation in the implementation stage is hardly any better; four out of every five managers in the survey reported that participation was restricted to either no involvement or "information" only, with less than 5% of managers stating that participation by employee representatives was characterised by "negotiation" or "joint decision-making". Thus, Italy is ranked 10th — only above Luxembourg and Portugal — in following the guidelines laid down by the joint opinion of Val Duchesse.

Employee representatives report an even bleaker picture; nine out of ten of them state that participation in the planning phase of new technology consists of either "no involvement" (48%) or "information only" (41%). Again, the degree of participation in the implementation phase shows little improvement; 86% report that participation consists of either "no involvement" (40%) or "information only" (46%).

How can this low degree of participation in technological change in Italy be accounted for? The most obvious explanation is that Italian unions have traditionally regarded collective bargaining as by far the most important way to govern labour-management relations and to protect the terms and conditions of their members in the workplace. Italian unions have shown an attitude of strong diffidence not only towards worker participation in the decision-making bodies of private companies, but also towards any form of bilateral institution set up to co-determine working conditions in conjunction with management (apart from in the Italian public service). Italy, unlike many other European Community countries such as Belgium, France, Germany, Greece, Luxembourg, the Netherlands, Portugal and Spain, has no works council system providing for a form of employee participation in management decision-making which has been laid down by legislation. Instead, Italian unions largely rely on a legal right (Art.19 of the 1970 Workers' Statute) to insert clauses into collective agreements which oblige Italian employers to inform the shop stewards' committees — factory councils (consigli di fabbrica) or councils of delegates (consigli dei delegati) of their investment plans and long-term business policies which affect the terms and conditions of Italian workers. Despite the absence of a works council provision by law in Italy, some would argue that the de facto position of the consigli, agreed with the management, is often much stronger in practice than is often supposed.

There is also a national fund, the Cassa Integrazione Guadagni, which, especially during the late 1970s, provided a degree of financial security for workers that, according to some Italian commentators, reduced pressure to negotiate new technology directly since this fund provided a cushion which workers could fall back on. In fact, it is important to stress that plant-level bargaining in Italy is relatively novel; it has to be seen against the longstanding importance not only of national collective bargaining but also of political deals. For example, the "Scotti" 1983 agreement provided certain benefits for unions/employees in exchange (partly) for union agreement to prohibit bargaining over issues that had already been resolved at a higher level. This should be taken into account when interpreting our survey results.

Thus, if we use our explanatory framework to account for the data in our survey, at least four of the factors can be said to be unfavourable in promoting employee involvement in technological change: management's reliance on its workforce to achieve its objectives for introducing the new technology; management style and its attitude to participation; legislative provisions which lay down participation rights for employees or their representatives on a range of matters at enterprise level and a decentralised industrial relations system. On the other hand, the low level of participation has to be seen against the broader political context of political and national deals. For example, the IRI protocol of 1984 between the three union confederations and the IRI groups of state-sector industries (which were not included in our survey) could be seen as a voluntary "problem-solving" agreement providing for extensive consultation on a wide range of issues including work organisation.[23]

Generally speaking, the diffusion of new technology in Italy is not as advanced as in other major European Community countries although this must be qualified by taking account of the North-South contrasts in industrialisation and the the level of prosperity. Most of the new technological applications have taken place in the banking sector, and there is little evidence to suggest that technological change in this sector has met with many industrial relations problems. The introduction of new technological innovations, particularly in Italian manufacturing industry, has not been dictated by competitive and product market factors requiring a great deal of employee commitment and co-operation in the workplace. As Colombo and Lanzavecchia claim in their 1983 study[24], technological innovation in Italy — at least in the early 1980s — was not creating particular problems of mobility and adjustment either in the labour market or in enterprises. In short, at the risk of over-generalisation, there does not seem to a great deal of dependence by management generally on employees to achieve its objectives by adopting new technology.

Another factor which has a negative impact on shaping the opportunities for participation in Italy is management style. Italian management has always viewed the political or "class" traditions of the trade unions with a deep suspicion, and their attitude to employee participation is questionable. Whilst the diffusion of new technology has accelerated in Italy during the 1980s, it largely affects white-collar workers, particularly in the services sector. In any case, compared with other major European Community countries, Italy has a relatively low percentage of its civilian work force employed in the services sector.

As we saw earlier, the trade unions in Italy have always accorded a primacy to collective bargaining as a means of rule determination in the workplace. There is no works council system which has been introduced by legislation.[25] Instead, Italian trade unions have preferred to defend their interests in Italian companies by subsuming technological change into the normal process of collective agreements. Working conditions in the workplace are negotiated at departmental level in such a way so as to impede unilateral imposition of managerial prerogative. In such circumstances, trade unions have invariably viewed participation generally with suspicion and sometimes with hostility.

One method that the Italian unions have employed to deal with technological change is to subsume it under the negotiation of agreements concerned with the organisation of work (Organizzazione del Lavoro: OdL). These agreements, negotiated at company level, rarely make explicit reference to new technology, but they do constitute a way of dealing with it. Such agreements sometimes cover matters such as harmonisation of conditions between blue-collar and white-collar employees, provide for increased flexibility in the organisation of tasks and generally cover means whereby work can be "humanised". Whilst this type of agreement is by no means common, it is nevertheless part of an increasing trend towards more company bargaining in Italy. Despite the emphasis placed on the need for centralisation in the past, plant bargaining is on the increase.

Another method that the Italian unions have used in order to bring about negotiations on new technology — particularly in large companies — is the use of the right to the disclosure of information on a wide range of company policy, ranging from information on investment programmes, technological change, subcontracting, the location of new sites and the likely consequences of technological change on employment and work organisation. Rights concerned with information disclosure began to find their way into national agreements in 1976-7.

Unfortunately, the unions have not been very successful in widening the scope of negotiating subject-matter to embrace these broad aspects of company policy. In many cases management imparts information to the unions on policy decisions that have already been taken at a time when it is impossible for the unions to influence company policy. It is rare for much consultation to take place and the unions usually end up in a very subordinate role vis-à-vis management. Management is very sceptical of the unions' competence to contribute very much of value to management in the consultation process, and they believe that union attitudes to change and modernisation in companies are still influenced by an out-dated stance which is rooted in the traditions of political or "class" ideology.

All this suggests that the gradual transition towards agreements which are more flexible in respect of work reorganisation in Italian companies is a very slow process indeed. Slow though this transition is, there are indications that Italian unions are viewing participation in a more favourable light and are increasingly less concerned with the protection of rigid rights in relation to jobs. Technological change in Italy is slowly but surely spawning changes in attitude among both employers and unions; both sides are gradually realising that they have an interest in a more flexible, participative and informal kind of bargaining at the workplace to deal with the issues thrown up by technological change.

This gradual change is reflected in the preferences of both sides for increased levels of participation in the future. The results for both managers and employee representatives in our survey indicate a decrease for "no involvement" in the planning phase of technological change from the past to the future (managers 37% to 24%; employee representatives 48% to 4%). The decrease for employee representatives is particularly striking. Whereas over half of all managers (53%) indicated that participation was limited to "information" in the past, 46% would prefer this kind of participation in the future. The percentage of managers indicating "consultation" in the past trebles from 8% to 23%. However, there is a large divergence of opinion between managers and employee representatives with respect to negotiation/joint decision-making in the planning phase; fewer managers want this kind of participation in the future (2% in the past and 7% for the future) than employee representatives (2% in the past and 32% for the future). This might well fore-shadow conflict between the two sides of Italian industry in the future about the type of participation that should exist in Italian enterprises.

In the implementation phase of new technology there is a similar picture. Taking managers alone, the percentage indicating "no involvement" in the past declines from 34% to 23% for their future preferences; there is also a shift in attitude for "consultation", with around twice the number of managers stating such a preference for the future (14% to 27%). Again, managers show a reluctance for "negotiation/joint decision-making" and the increase here is marginal (4% to 9%).

The preferences of employee representatives in the implementation phase again indicates a wide divergence of opinion between their management counterparts; four out of ten employee representatives indicated "no involvement" in the past and this figure shrinks to less than one in 20 for the future. Just under half of all employee representatives stated that participation was characterised by "information only" and only one in ten claimed that there was any consultation. Italian employee representatives show a clear preference for negotiation/joint decision-making in the future; only 2% claimed that such a form of participation existed in the past whereas seven in 20 would like this in the future. Italian employee representatives are generally in line with their European counterparts in their support for ETUC policies of advocating negotiation and/or joint decision-making as the basis for the introduction of technological change in the future; 32% in the planning phase and 35% in the implementation phase.

In sum, the traditional antagonistic relations between the employers and trade unions in Italy are reflected in the results of our survey. While there is a gradual shift in attitude taking place on both sides, it is clear that there is a long way to go before a higher degree of trust can exist between the two sides of Italian industry. The trade unions are still deeply suspicious of embarking on the road to participation in technological change, preferring instead to adopt a defensive stance by seeking to restrict the issues of technological change under the umbrella of collective bargaining.

Managers too can see no reason why participation in technological change should extend beyond mere "information"; seven out of every ten Italian managers are not prepared to envisage anything more than the supply of information on their intentions for the introduction of technology into the workplace. Italian managers are the least willing of all their European Community counterparts — with the exception of Portugal — to contemplate consulting with their workforce on technological change. This figure is well below that envisaged by the Val Duchesse social dialogue joint opinion between the ETUC and UNICE.

The influence of our explanatory factors in relation to the levels of participation in technological change in Italy is summarised in Table 8.10 below:

TABLE 8.10

Variable	
Technological Objectives	Little dependence by management on the skills and problem-solving abilities of employees but gradually increasing; management jealously guards its prerogative and is sceptical of the competence of employees to offer much of value in technological change but this attitude is gradually but slowly changing; diffusion of new technology lower than most EC countries
Management Style	Views political or "class" traditions of Italian unions with deep suspicion; hostile to improved levels of participation but this is gradually changing; low trust relations with employees
Bargaining Power	Trade unions organised in three main confederations on political lines in CGIL, CISL and UIL; inter-union rivalry but recent attempts at achieving working unity; "class" traditions; moderate trade union density (40%); importance of shop steward committees (consigli di fabbrica); high premium placed on collective bargaining rather than participation or co-determination but this is gradually changing; bargaining power generally weak but pockets of strength in certain sectors; plant-bargaining relatively novel
Regulation	Regulation primarily through collective bargaining; no legislation providing for institutionalised works councils but rights emanating from Workers' Statute (Statuto dei lavoratori); recent political accords e.g. IRI protocol of 1984

Industrial Relations System	Very decentralised; collective bargaining largely at national level but plant-level bargaining on increase; pace of change in Italian industrial relations slow

11. LUXEMBOURG

The small size of Luxembourg and the narrow range of its economic base have greatly influenced its pattern of industrial relations. The Luxembourg model of worker participation in much less formal than the models found in neighbouring European Community countries, and it aims to establish worker representation mechanisms which are tailored to the size and nature of the enterprise.

The influence of trade unions in the Grand Duchy is felt most keenly in the coal and steel industries, which are important to the economy. There are two major trade union confederations: the socialist CGT (Confédératione général du Travail) and the Christian LCGB (Lëtzebuerger Chrëschtleche Gewerkschafts-Bond) which together have approximately 64,000 members.

The major components of the Luxembourg model of worker participation are hybrid forms of other systems. The personnel delegates (délégués du personnel) perform a function that combines some of the duties of a works councillor with the grievance-handling tasks of a shop steward. The mixed-enterprise committee (comité mixte) in a small firm is analogous to a works council, and in a large firm, it is similar to a supervisory board.

Collective bargaining is carried out entirely at the level of the enterprise, or, in the case of multi-establishment companies, at company level. Employers must bargain if called upon to do so; if they refuse there is provision for conciliation.

Luxembourg has the lowest percentage of managers of all the European Community countries who report participation levels in the planning stage of new technology in compliance with the Val Duchesse joint opinion guidelines. Our survey results for Luxembourg show that 19 out of 20 managers state that employee involvement in the planning phase of technological change extends to nothing more than information. This statement is reflected by a similar percentage of employee representatives. In the implementation phase the level of employee involvement is not much higher; 93% of managers report that it consists of either no involvement or the mere provision of information. Again, 92% of employee representatives confirm this view.

How can we account for this very low level of participation in Luxembourg? After all, Luxembourg has had a system of worker participation since 1974, and we would expect employee involvement to be much higher.

Since 1979 a law has been in operation which requires the establishment of a committee of personnel delegates (délégués du personnel) in every private sector undertaking with more than 15 employees, the number of delegates depending on the number of employees. Candidates are elected from among employees on a proportional representation basis by secret ballot and the trade unions present lists of their candidates to voters. However, the functions of these délégués du personnel are very limited, and only extend to conciliation duties in the event of disputes, representation of employees' interests to management, participation in the management of social facilities in the enterprise and in the training

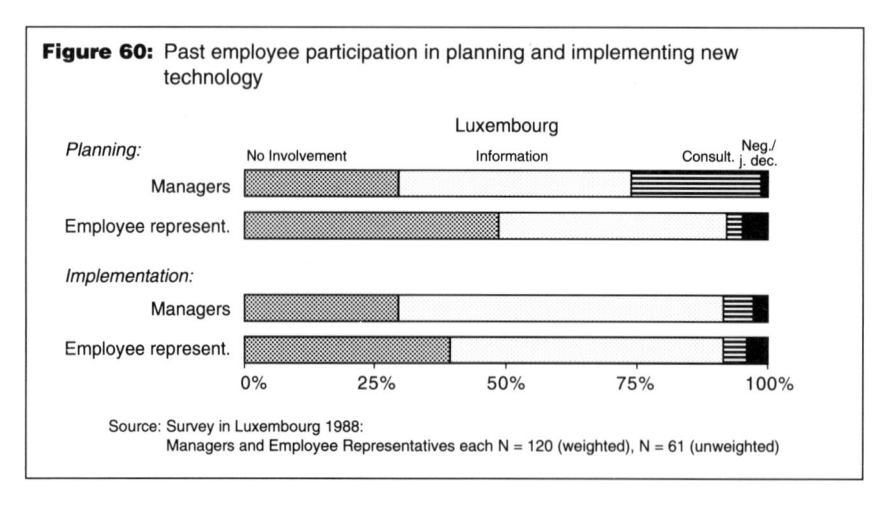

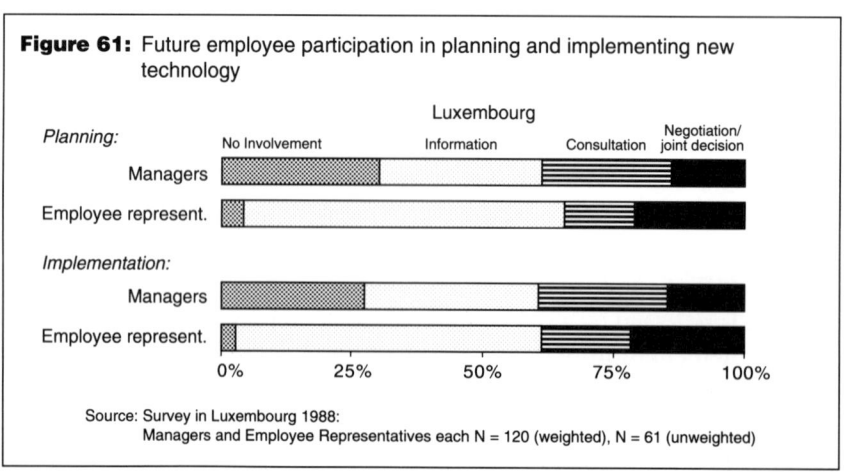

Figure 60: Past employee participation in planning and implementing new technology

Luxembourg

Source: Survey in Luxembourg 1988:
Managers and Employee Representatives each N = 120 (weighted), N = 61 (unweighted)

Figure 61: Future employee participation in planning and implementing new technology

Luxembourg

Source: Survey in Luxembourg 1988:
Managers and Employee Representatives each N = 120 (weighted), N = 61 (unweighted)

of apprentices, dealing with accident prevention and the right to give its opinion on working regulations.

The functions of the committee of délégués du personnel ends where that of the comité mixte d'enterprise starts, and the two bodies do co-operate together. The comités mixtes d'enterprise have existed since 1974 and must be set up in all private sector enterprises employing more than 150 workers. They consist of an equal number of employer and workers' representatives and the size of this body increases according to the number of employees in the enterprise. As in the case of the délégués du personnel, employee members of the comité mixte d'enterprise are elected by proportional representation from among the staff. This body is chaired by the managing director and a secretary is designated from among the employee representatives.

What are the functions of the comité mixte d'enterprise in Luxembourg? It can take decisions on health and safety, general assessment criteria concerning the selection, recruitment, change of classification or dismissal of workers. More importantly for our purposes, it has a right to information and consultation[26] on the construction, transformation and the extension of production units, new working methods, the effects of economic and financial decisions on the level of employment, and economic and financial trends.

Given the functions of the comité mixte d'enterprise in Luxembourg outlined above, it is particularly puzzling why there is not more consultation shown in our results. The trade unions, regardless of political allegiance, are known to be very dissatisfied with the present law, and they are pressing for a greater degree of co-determination.

What preferences do the two parties have for the future? Managers seem to envisage a greater level of involvement by employee representatives in the future, and seem to have found the experience of participation useful; two out of five managers are prepared to countenance consultation or negotiation/joint decision-making in the planning phase of technological change and a similar number are prepared for similar levels of employee involvement in the implementation phase.

Surprisingly, employee representatives indicate that they are happy with a level of participation in the planning phase which extends at best to no more than information; 66% of them had this view. In a similar vein, three out of five employee representatives seem content with no more than information in the implementation phase of technological change. Clearly, there seems to be a conflict of opinion between the trade union leadership and its rank-and-file membership in Luxembourg enterprises. Less than two per cent of Luxembourg employee representatives appear to support ETUC policies of influencing technological change by negotiation or joint-decision-making —by far the lowest of all their counterparts in other EC countries.

It should come as no surprise that the results for Luxembourg fall well short of the Val Duchesse joint opinion between the ETUC and UNICE on information and consultation in new technological introduction. Both managers and trade union members appear not to be concerned with the need for more employee involvement in the light of Val Duchesse. The survey results for Luxembourg fall well below the European average in our survey. The influence of our explanatory factors are summarised in Table 8.11 below:

TABLE 8.11

Variable

Technological Objectives	Little dependence by management on the skills and problem-solving abilities of employees; few large enterprises
Management Style	Co-operative
Bargaining Power	Trade unions weak; inter-union rivalry between the CGT and LCGB; trade union strength restricted to declining heavy industries; reliance on enterprise bargaining
Regulation	Works council legislation provision since 1974 but gives little influence to employee representatives; collective bargaining entirely at enterprise level
Industrial Relations System	Extremely decentralised

12. PORTUGAL

The current Portugese industrial relations system is one of the youngest in the European Community, dating from the 1974 revolution which ended the previous dictatorship and introduced democracy. The current system still retains elements of the pre-revolutionary system, especially in the high degree of state regulation. During the years of the Salazar dictatorship, collective bargaining, as such, did not exist in Portugal, but following the 1974 revolution the practice of free collective bargaining was introduced. The system which has since developed has been marked both by features of the pre-revolutionary set-up — particularly a high degree of state intervention both through legislation and administration — and by the long period of economic crisis which Portugal underwent from 1976 to 1984.

The all pervasive influence of the state on the conduct of industrial relations extends into the way trade unions and employers are organised. Both must be registered with the Ministry of Labour in order to enter into collective agreements. There are strict standards imposed on trade union constitutions. Legislation also establishes the division of responsibilities between unions and enterprise-level works councils (Commisoes de Trabalhhadores): the former have sole responsibility for collective bargaining and the declaration of strikes (except in the rare cases where the majority of employees are not union members), while the latter have, almost exclusively, rights to information, consultation and participation.

The law also stipulates collective bargaining procedure, content and form. All negotiations have fixed time-limits for each phase; specific items of information for collective bargaining must be supplied; and once concluded, the agreements are then registered with the Ministry and are legally binding. The content of bargaining is restricted to a number of subjects.

Portugese law provides for three types of collective agreement: the Collective Labour Contract or Contrato Colectivo de Trabalho (CCT), which is negotiated between one or more trade unions and one or more employers' associations at sector level; the Collective Labour Agreement Acordo Colectivo de Trabalho (ACT), negotiated at company level between unions and more than one employer (but not an employers' association); and a Company Agreement Acordo de Empresa (AE), negotiated between one or more trade unions and a single employer. Since 1976, there has been a marked trend away from local/regional bargaining towards national bargaining, and away from "horizontal" agreements — covering a particular occupation or grade in a number of sectors — towards agreements covering various groups of workers within an individual sector. A survey of 271 collective agreements published in 1987[27], revealed that 182 of them were national in scope, while the number of "horizontal" agreements had declined to 66 from 100 in 1983. The coverage of collective agreements among the Portugese work-force is wide; figures released by the Portugese Ministry of Labour Statistics Office show that, in 1986, 79.4% of the labour force was covered by a Çollective Labour Contract, or CCT.

Most of the content of collective agreements in Portugal concerns pay, and for many years wage bargaining has been strongly influenced by "wage ceilings" (tectos salarais) set by the government at the beginning of each year; since 1986 central pay norms have when possible been set by a tri-partite Standing Committee on Social Consultation (conselho permanente de concertaçao sociale). However, the pay norms and the ceilings are based on inflation projections which invariably prove to be well below the actual rate; as a result the wage levels fixed by agreement tend to bear little relation to actual earnings.

The parties in the Portugese industrial relations system are organised in a fragmented manner, and collective bargaining is conducted both by formal groupings on the two sides — usually federations — and by ad hoc groupings of varying size. The wider political and social aspects of employers' and employees' interests are centralised in the three employers' confederations CIP, CCP and CAP — representing industry, commerce and agriculture respectively — and the two union confederations CGTP-IN and UGT — oriented towards the Communists and Socialists respectively. Relations between the two main union confederations were strained until recently, but now the CGTP-IN has joined the Standing Committee. Similarly, there are indications that the three main employers' groupings may join together in the near future.

How do the Portugese works councils operate? The role of works councils is again strictly regulated by law. They are composed entirely of employee representatives and are elected by secret ballot for one- or two-year terms by proportional representation. Frequently, trade union members will dominate the composition of the councils but this is not always the case. The size of the works councils will vary in relation to the size of the number of people employed in a particular establishment. Managers are required to meet the works council at least once a month.

The rights that works councils have are specifically laid down by law. They have the right to information[28] on "the implications of production plans for the use of labour and equipment" but no legal right to be consulted on matters other than dismissals, redundancy, holidays, working hours, job classification and company location. In state-owned companies they have the additional right to be consulted on changes to the company statutes and in the appointment of management.

In addition, workers in state-owned companies have the right to elect at least one representative to sit on the management board of their company, but the unions have long complained about the ineffectiveness of this measure in practice.

The Portugese results are by far the worst of all European Community countries; clearly, Portugese representatives have very little influence on technological change. Moreover, there is a great deal of agreement between managers and employee representatives about this very low level of participation. In the planning phase of new technology introduction, 17 out of every 20 managers state that there is no involvement whatsoever; almost the exact percentage of employee representatives agree with this view. Two out of every 20 managers and employee representatives maintain that technological change in the planning phase is characterised by information only.

In the implementation phase there is again a strong consensus between the two parties. 18 out of every 20 managers and employee representatives interviewed stated that this phase involved information about technological change. This figure is the highest of all the European Community countries. One out of 20 managers and employee representatives interviewed claimed that technological change is characterised by negotiation or joint decision-making.

The Portugese results stand out from every other European Community

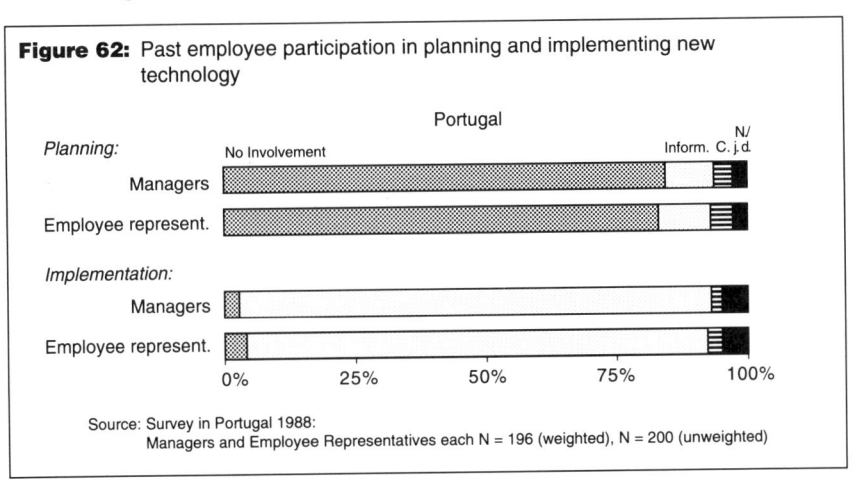

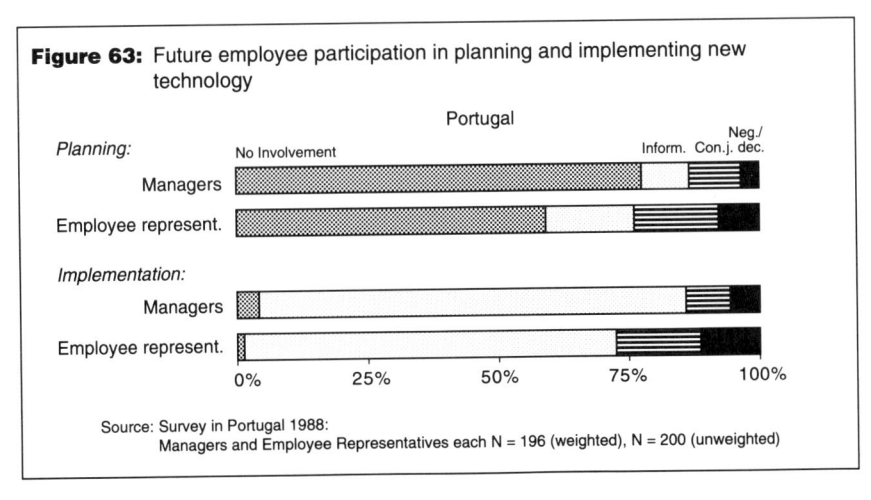

Figure 62: Past employee participation in planning and implementing new technology

Source: Survey in Portugal 1988:
Managers and Employee Representatives each N = 196 (weighted), N = 200 (unweighted)

Figure 63: Future employee participation in planning and implementing new technology

Source: Survey in Portugal 1988:
Managers and Employee Representatives each N = 196 (weighted), N = 200 (unweighted)

country, not only in the abnormally high level of no involvement by employee representatives in the planning stage of new technology but also in abnormally high reliance on information provision in the implementation phase of new technology. Clearly, this can only explained by reference to what legislation requires. A detailed examination of the legislation operative in Portugal which was set out in a document produced by the European Trade Union Confederation[29] confirms this conclusion; virtually every facet of the Portugese industrial relations system is legally defined down to the smallest detail. Works councils must by law be given information on the implications of technological change in the implementation phase but there is no provision to be given any information about wider company strategy generally. Our results therefore reflect the legal strictures which define what can and cannot be the subject of participation insofar as technological change is concerned. The subjects on which employee representatives are consulted do not include technological change, either in the planning or implementation phase.

If we compare the Portugese results with the guidelines laid down in the Val Duchesse joint opinion, they show that only 6% of Portugese managers report that participation levels in both the planning and implementation stages of new technology involve consultation or higher levels. Both sides of Portugese industry agree on this. Compared with most other European Community countries, technological advancement is poorly developed in Portugal, and it seems that employee participation in technological change is not yet a major issue between employers and unions. Portugal's industrial structure is extremely fragmented, with a large number of small enterprises and relatively few large undertakings, principally multinational companies and industries nationalised in 1975. This fragmented industrial structure is reflected in the way in which employers and employees are represented.

The heavy reliance of both parties on legislation means that the current industrial relations system in Portugal is overly restrictive and inflexible, with an excessively high level of government involvement. In such circumstances, the Portugese industrial relations system is incapable of accommodating technological and organisational change.

What preferences do managers and employee representatives interviewed in our survey have for the future? Again, our results show that Portugal is way out of line with the rest of the European Community. Just over three quarters of managers want no involvement in the planning stage and so do three out of five employee representatives. There is also a consensus between both sides in the implementation stage, with over four out of five managers preferring information and seven out of ten employee representatives wanting information. It goes without saying that past practice in participation in technological change is well below that envisaged by the Val Duchesse joint opinion between UNICE and the ETUC. Unfortunately, the views of both managers and employee representatives in relation to their preferences for levels of participation in the future hold out little prospect of much improvement in the level of involvement. It ought to be mentioned, however, that the major political parties in Portugal are aware of the need to change the way in which collective bargaining is conducted.

The influence of our explanatory factors on the levels of participation in Portugal is outlined in Table 8.12 below:

TABLE 8.12

Variable

Technological Objectives	Little dependence by management on the skills and problem-solving abilities of employees; technology poorly developed; technological change not yet a major issue between both sides of industry; few large undertakings; legacy of Salazar dictatorship still evident
Management Style	Hostile to participation; no involvement by employees almost universal
Bargaining Power	Trade unions strictly regulated by law; weak bargaining power; emphasis on pay bargaining; collective bargaining mainly at national level
Regulation	Regulation almost entirely by law down to the finest detail; wide coverage of collective bargaining; inflexible and restrictive legal system; works councils operation strictly regulated by law
Industrial Relations System	Centralised but little legal powers given to employees at establishment level on technical change; fragmented organisation of unions (CGTP-IN and UGT) and employers

---

NOTES

1. There has been extensive legislation in the public sector in Ireland, but our survey did not cover the industries concerned.

2. European Foundation for the Improvement of Living and Working Conditions, 'Co-operation and Workers' Participation in Danish Factories', Dublin, 1981.

3. There are four trade union confederations: the DGB, with 17 affiliated industrial unions, which is by far the largest and the most influential; the DBB, which organises Civil Service Officials; the DAG, a small white-collar confederation; and the CGB, the Christian Trade Unions of Germany, which operates in some regions. Trade union density in Germany is around 38%. The real power within the DGB lies with the single industrial unions, the largest and most influential of which are IG Metall (2,535,000 members - 1988), the Union of Public Service, Transport and Communication Workers (1,173,000 members - 1988), and IG Chemie-Papier-Keramik (635,000 members - 1988).

4. Statistisches Bunbesamt, 1985.

5. Rudi Schmidt and Rainer Trinczek, 'Does the 38.5 hour week collective agreement change the West German system of co-determination?', in C.J. Lammers and G. Széll, International Handbook of Participation in Organizations, Vol. I, Oxford University Press, 1989, p. 296.

6. Paul Cullen, 'The worker participation question', in Hugh Pollack, *Reform in Industrial Relations*, , The O'Brien Press, Dublin, 1982, pp. 82-89.

7. Joe Wallace, 'New information technology and participation in Ireland and the European Community', College of Business, University of Limerick, November 1989.

8. *European Industrial Relations Review*, 186, July 1989.

9. A. Teulings, 'A Political Bargaining Theory of Co-determination', *Organisation Studies*, 8, 1987

10. *De OR en zijn bevoegdheden*, Instituut voor Toegepaste Sociologie, 1986; reported in *European Relations Review*, 150, July 1986.

11. S.C. Looise, *Werknemersvertegenwoordiging op de Tweesprong*, Alphen A/O Rijn, 1989.

12. See Waut Buitelaar, *Technology and Work*, Gower 1988.

13. Michael Albertijn et al., 'Technology agreements and industrial relations in Belgium', *New Technology Work and Employment*, 5, 1, 1990.

14. Jones, B. 'Work and flexible automation in Britain: a review of developments and possibilities', Work, Employment and Society, 2, 4, 1988.

15. Christel Lane, Management and Labour in Europe, Edward Elgar, 1989, pp. 189-195.

16. Bessant, J.R. and Grunt, M., Management and Manufacturing Innovation in the United Kingdom and West Germany, Gower, 1985.

17. See Michéle Tallard in R. Hyman and W. Streeck, New Technology and Industrial Relations, Basil Blackwell, Oxford, 1988, pp. 290-292.

18. European Industrial Relations Review, 179, December 1988.

19. Reported in European Industrial Relations Review, July 1989.

20. Z. Tzannatos (ed.), Socialism in Greece: the first four years, Gower, 1986.

21. P. Fakiolas, Factors determining industrial employment in Greece, (in Greek) cited in 'Equal Pay in Greece and Britain', Industrial Relations Journal, 18, 4, 1987.

22. Yearbook of Labour Statistics, 1983, ILO, Geneva.

23. See the contrast between industrial relations in Italian state-owned companies and private sector companies in P. Garonna and E. Pisani 'Il italiano nella transizione: la crisi del sindicalismo politico' in R. Edwards, P. Garonna and E. Pisani (eds.), Il Sindicato oltre la Crisi 1984.

24. Colombo, U. and Lanzavecchia, G., La Posizone relativa alla tecnologia italiana, Paper presented to the Seminar Nomisa on Technological Innovation and the structure of production: the Italian situation, Milan, 1983.

25. Note however, the rights given to workers as a result of Art. 19 of the 1970 Workers' Statute, mentioned earlier.

26. European Trade Union Confederation, Mobilisation Conference, Europe's future: strong representation of workers' interests in the undertakings of Europe, Ostend 16-17 October 1989.

27. Boletim do Trabalto e Emprego (Labour and Employment Gazette, Ministry of Employment) 1987

28. European Trade Union Confederation, Mobilisation Conference, Europe's future: strong representation of workers' interests in the undertakings of Europe, Ostend 16-17 October 1989

29. European Trade Union Confederation, Mobilisation Conference, Europe's future: strong representation of workers' interests in the undertakings of Europe, Ostend 16-17 October 1989

Chapter Nine

Summary of Individual Country Results

1. Introduction

Chapter VIII of this report gave a detailed account of the levels of participation in existence in all countries of the European Community in relation to technological change in enterprises. It sought to explain the great diversity that was found from one country to another by using an explanatory framework which consisted of five major factors which shaped the opportunities for employee participation in technological change. These factors were based on those which had been identified from the growing literature on new technology and its relationship to work organisation, skills, employment levels, work patterns from the mid-1970s onwards. Our survey data also gave an indication of what the preferences of both sides of industry were in each country as to the type of participation they envisaged for the future, based on their past experience.

In this chapter we review the results within a broader political context and evaluate the great diversity in participation practice that was found throughout the European Community to see if any distinct patterns emerge from our results. A cursory reading of Chapter VIII indicates that the aggregate results at a European-level conceal an immense variation throughout the Community not only in terms of participation practice but also in relation to the conduct of industrial relations. There are considerable variations from one country to another in the levels of no involvement whatsoever by employee representatives in the introduction of new technology in their enterprise; the degree of information provision; the amount of consultation that takes place; the level of negotiation between both sides of industry and in the amount of joint decision-making or co-determination. There are also variations in the opinions of respondents from both managers and employee representatives on the level of participation in existence and on the kind of participation that they would prefer in the future.

2. A comparative evaluation of the levels of participation in the Member States of the European Community

Perhaps the most obvious way of evaluating the results is to examine the ranking of all 12 countries in terms of the level of participation in existence at the time the interviews took place in 1987-88. As we saw in Chapter VIII, Figures 64 and 65 gave a ranking of countries indicating the pattern of participation reported by managers and employee representatives respectively in the **planning** phase of the introduction of new technology. We ranked the order of the countries in the Figures so that those with the greatest degree of negotiation or joint decision-making would appear at the top and those with the least degree of this higher-level of

Summary of Individual Country Results

Figure 64: Past participation in planning for new technology in the EC countries - Managers

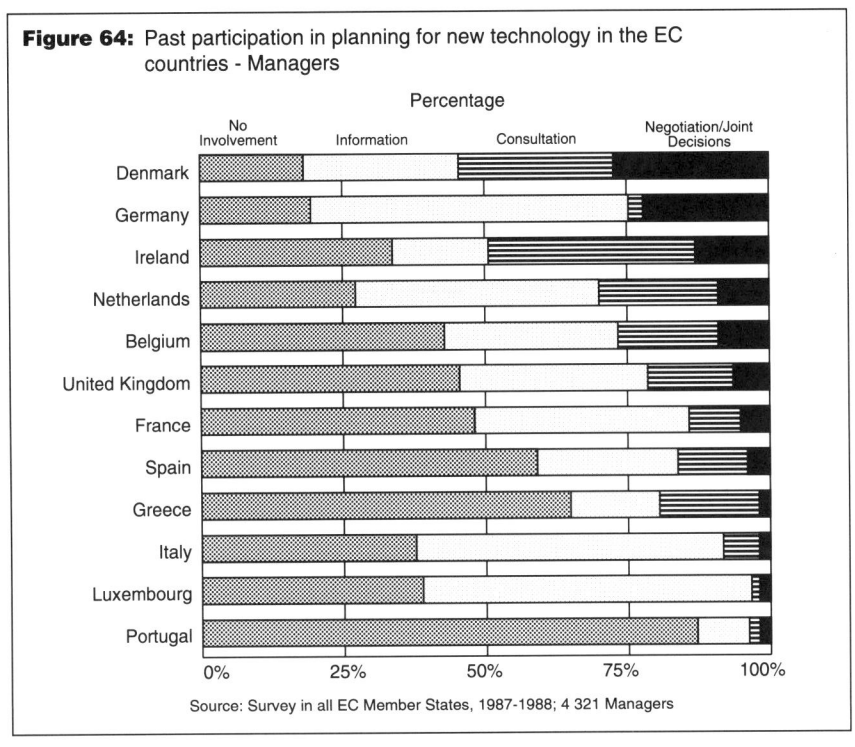

Percentage

Categories: No Involvement | Information | Consultation | Negotiation/Joint Decisions

Countries (top to bottom): Denmark, Germany, Ireland, Netherlands, Belgium, United Kingdom, France, Spain, Greece, Italy, Luxembourg, Portugal

Source: Survey in all EC Member States, 1987-1988; 4 321 Managers

Figure 65: Past participation in planning for new technology in the EC countries - Employee representatives

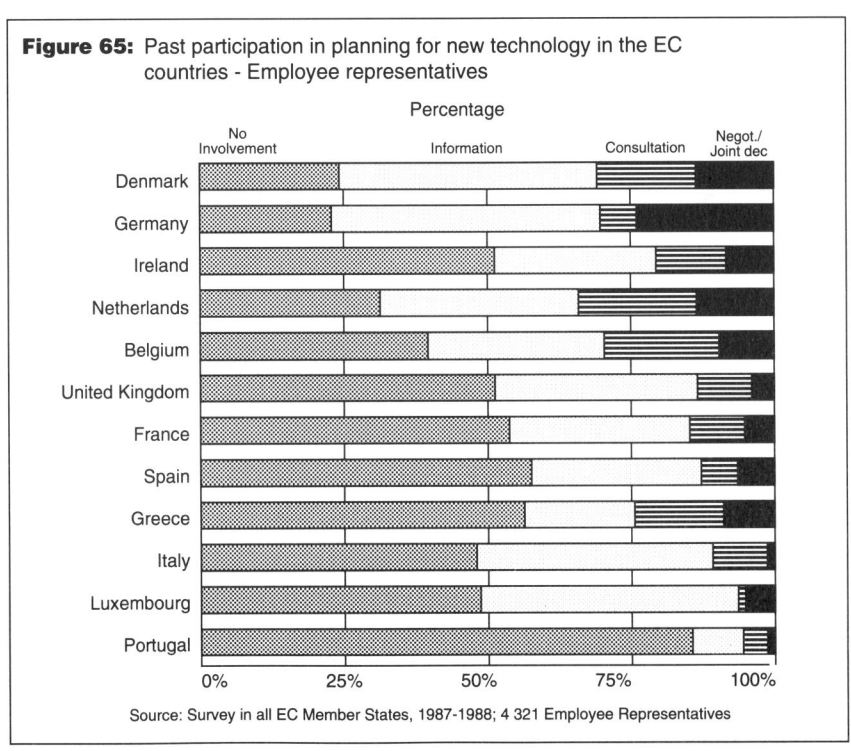

Percentage

Categories: No Involvement | Information | Consultation | Negot./Joint dec

Countries (top to bottom): Denmark, Germany, Ireland, Netherlands, Belgium, United Kingdom, France, Spain, Greece, Italy, Luxembourg, Portugal

Source: Survey in all EC Member States, 1987-1988; 4 321 Employee Representatives

participation at the bottom. Not surprisingly, there is some difference in the responses of both sides, the magnitude of which varies from one country to another; in some cases employee representatives are more sceptical than their management counterparts (Figure 65), but in other cases they are less so. Nevertheless, a distinct pattern can be seen.

Two countries, **Denmark** and **Germany** clearly have the highest level of negotiation or co-determination. There is also a middle-ranking group of countries which have higher levels of negotiation or co-determination: **Ireland**, the **Netherlands** and **Belgium**. At the bottom of this ranking there are a large number of countries where the amount of negotiation or co-determination is minimal: **United Kingdom, France, Spain, Greece, Italy, Luxembourg** and **Portugal**.

There is a somewhat similar ranking pattern among these countries if we examine the degree of no involvement reported by both sides in the planning phase of new technology. Again, **Denmark** and **Germany** have the least amount of no involvement of all the countries in the survey. These are followed (in order) by a middle-ranking group of countries including the **Netherlands, Ireland, Italy,** and **Luxembourg** followed by a bottom-ranking cluster of countries which include **Belgium, United Kingdom, France, Spain** and **Greece**. At the very bottom of our table comes **Portugal**, a country which stands out prominently in our survey data. As we mentioned earlier, this country's unusual position can be largely explained by the way in which the law there lays down very detailed and precise provisions which define not only the form, content and conduct of collective bargaining but also the rights for information and consultation relating to Portugese works councils.

We would also expect a similar ranking of countries in the **implementation** phase of technological change; at this phase the influence of our explanatory factors still operates. Figures 66 and 67 show a similar pattern to the planning phase. If we group the countries by ranking them in terms of their degree of negotiation or co-determination a similar pattern emerges. **Denmark** still features at the top of the table closely followed by **Germany**; there is a middle-ranking group of countries which includes **Ireland**, the **Netherlands, United Kingdom, Greece, France** and **Spain**. At the bottom of the table there is a cluster of three countries with a minimal level of negotiation or co-determination: **Portugal, Italy** and **Luxembourg**.

If we rank all the countries in our survey according to the level of no involvement by employee representatives in the implementation phase of technical change we encounter some surprising variations. **Portugal** has the lowest level of no involvement followed by the **United Kingdom**. There is then a middle-ranking group of countries which includes

Summary of Individual Country Results

Figure 66: Past participation in implementing new technology in the EC countries - Managers

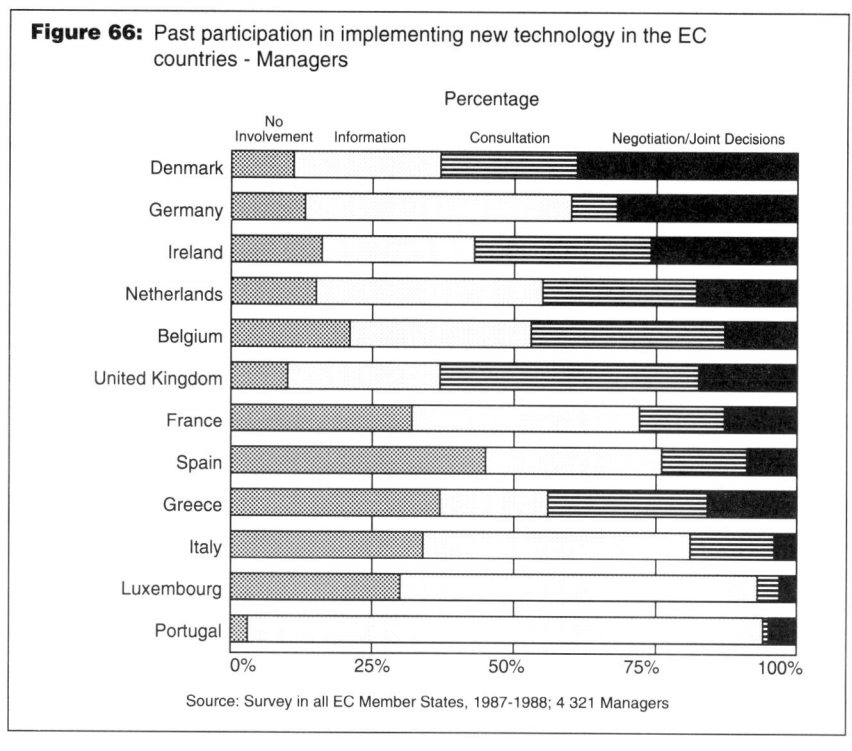

Source: Survey in all EC Member States, 1987-1988; 4 321 Managers

Figure 67: Past participation in technology implementation in the EC countries - Employee representatives

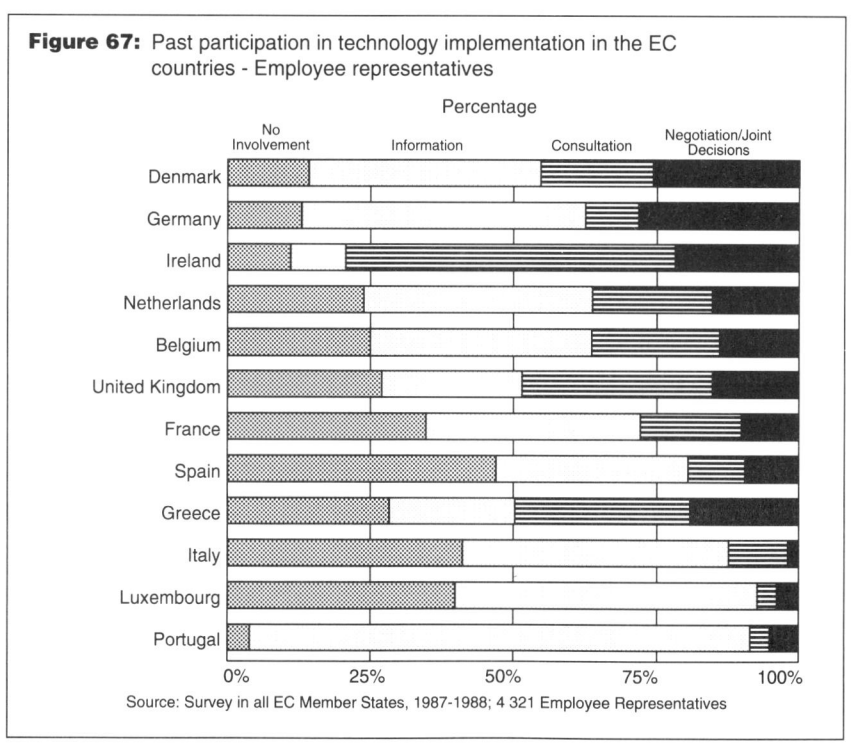

Source: Survey in all EC Member States, 1987-1988; 4 321 Employee Representatives

Denmark, the **Netherlands, Ireland** and **Belgium**; at the bottom of the table we find **Luxembourg, France, Italy, Greece** and **Spain**.

3. Using our explanatory factors to account for differences in participation levels

We can safely assume that the figures given above give a reasonably accurate picture of the state of participation that existed in all the European Community countries at the time our interviews took place in 1987-88. How can we use our **explanatory factors** to account for these rankings in both the planning and implementation phases? If our explanatory factors are of any value we would expect to find that nearly all the factors are favourable in shaping the opportunities for participation in the higher-ranking countries with strong forms of participation such as **Denmark** and **Germany**; similarly, in the case of those countries with the lowest levels of participation we would expect that many of our explanatory factors to be unfavourable (e.g. **France, Spain, Greece, Italy, Luxembourg** and **Portugal**).

If we consider the case of **Denmark** first, all five of our factors are very favourable. There is a high dependence by management on the skills and problem-solving abilities of employees; management is willing to negotiate over a wide range of technological issues and managers could be said to adopt a "Scandinavian"-type, co-operative management style where the unions are accepted as useful partners in making the fullest use of new technology. The Danish trade unions in turn are committed to technological change and have a very high density; whilst the unions are determined to secure the best deal possible in negotiating with management over technological change, there is a commitment to strive for agreement in the interests of both parties. The Co-operation Committees in Denmark work well, and they produce a substantial degree of co-determination on work organisation issues quite separate from the system of national bargaining. There is also a system of electing employees on to the Boards of Danish companies. All this is buttressed by a highly centralised industrial relations system where both employers and trade unions are able to exert authority over their respective affiliates; both parties cover the entire labour market.

All five factors can be said to be largely favourable in **Germany** as well although not to the same extent; there are crucial differences between the two countries. First, the German industrial relations system is much more legalistic, and whilst legislation there provides a very workable and co-operative co-determination system through its works councils, it is only centralised at sectoral level. Secondly, management style is much more diverse from one company to another; whilst it is generally co-operative, particularly in the larger companies, it tends to be paternalistic in smaller enterprises. Moreover, German managers are known to oppose further improvements in works council legislation which would have the effect of giving more influence to employee representatives in new technology. Finally, whilst German trade unions are well-resourced and well-organised, there is an absence of a formal trade union presence at the workplace where major decisions on new technology are taken.

It is a much more complex task to summarise the influence our factors in the case of the group of middle-ranking countries i.e. **Ireland,** the **Netherlands** and **Belgium**. Here we are faced with a mixture of favourable and unfavourable factors. We must interpret the results of **Ireland** with some caution because of the small sample of respondents (see Chapter VIII). Most of the factors which relate to this country are generally unfavourable, although management there is receptive to the outcome of "National Understandings" between themselves, the government and the trade unions. Whilst there is little dependence by Irish management on the skills and problem-solving abilities of employees, trade unions have a significant degree of bargaining power and the industrial relations system is gradually becoming more centralised. However, there is no legislation to promote participation, except in state-owned industries - which were in any case not included in our survey.

In the **Netherlands** all five of our factors are favourable and it might be asked why, with a similar co-determination system as in **Germany** this country does not have such a high level of participation as its neighbour? This is possibly because the trade unions there are weaker and are divided and the links that the unions have with the works councils are somewhat tenuous. In addition, the legislative provisions on works councils in Germany give greater co-determination rights than in the Netherlands.

All our factors are also generally favourable in **Belgium**, but the works council system in this country has a limited role in enabling employees to influence technological change because it is essentially only advisory. Despite the negotiation of a national technology agreement there in 1983, the evidence suggests that this agreement is not always used and Belgian workplace representatives often use other legal provisions or technology clauses are included in collective agreements.

As we move down the table and apply our explanatory factors to the lower-ranking countries, the factors become more and more unfavourable. In descending order these countries are the **United Kingdom, France, Spain, Greece, Italy, Luxembourg** and **Portugal**. If we consider each factor in turn we find that only in **France** and the **United Kingdom** is there a dependence by management of the skills and problem-solving abilities of the labour force; this explains the comparatively high level of consultation in Britain in the implementation stage of technological change.

Management style in all these low-ranking countries is generally unfavourable; in the **United Kingdom** there is evidence of a shift away from negotiation towards more consultation during the last decade and management are becoming increasingly paternalistic in their style. However, this serves only to increase the opportunities for employee representatives to be involved at the implementation stage of technological change. In **France** the Auroux reforms are working very slowly, and French management still jealously guards its managerial prerogative. In countries such as **Spain, Greece** and **Portugal** there is a legacy of a dictatorial past and management in these three countries are opposed to participation legislation which has been introduced by the new democratic governments, and the trade unions are weak, poorly-resourced and inexperienced. Apart from the **United Kingdom** and **Greece**, all these low-ranking countries have trade unions which are divided on political or religious lines, and even where there have been recent efforts within the wider trade union movement to forge closer links and promote a greater degree of unity (e.g. **Spain, Italy** and to a lesser extent **France**), such efforts have had very little impact as yet on participation. In **Greece** there has been a recent split in the GSEE.

Whilst all these low-ranking countries (with the exception of the **United Kingdom** and **Italy**) have various forms of formal, institutional works council provisions which have emanated from legislation, in many of these countries the trade unions eschew participation and have a greater preference for subsuming technological change under collective bargaining. In any case, the works council systems in many of these countries are inadequate in providing much influence to employee representatives in dealing with issues associated with new technology, or the links between the trade unions and the works councils themselves are tenuous. None of these countries have centralised industrial relations systems.

Our explanatory factors therefore, have proved to be useful in explaining the vast differences that exist between countries across the European Community. As we emphasise later, it is clear that the considerable differences in levels of participation that exist from one country to another in the European Community are largely shaped by the historical and cultural features of the industrial relations system in each particular country.

Summary of Individual Country Results

Figure 68: Past participation in planning for new technology in the EC countries - Managers

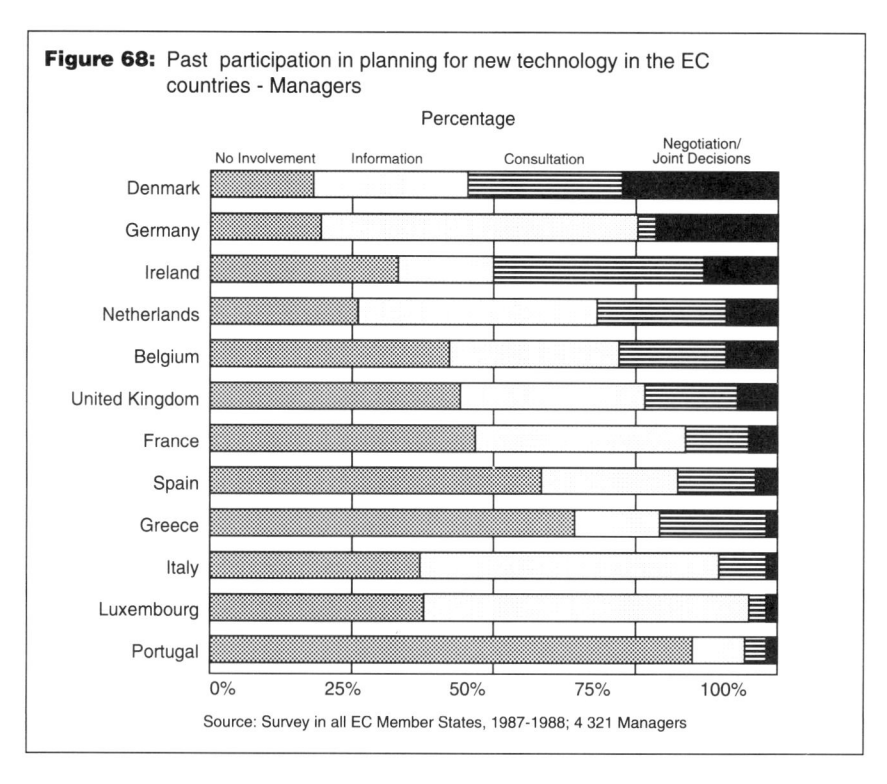

Source: Survey in all EC Member States, 1987-1988; 4 321 Managers

Figure 69: Future participation in planning for new technology in the EC countries - Managers

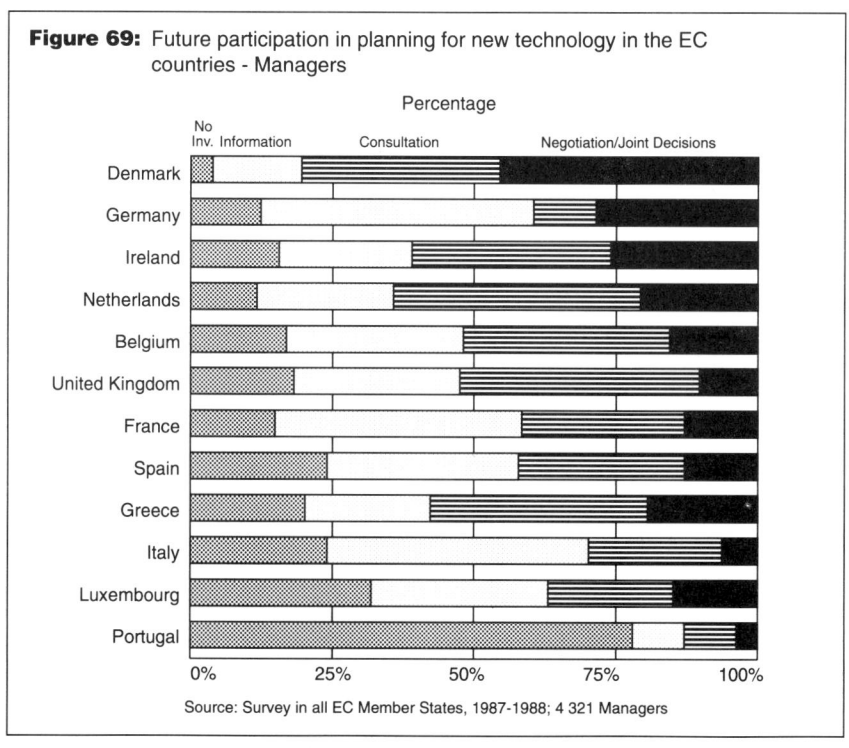

Source: Survey in all EC Member States, 1987-1988; 4 321 Managers

4. Assessing the potential for participation in the future

One of the most striking aspects of our survey is that in **every** European Community country both sides indicated in their responses that they wanted greater levels of participation in the future. Not only did this mean that the level of no involvement decreased substantially in the preferences of both sides for the future compared to what existed at the time of our interviews in 1987-88, but also both sides envisaged more negotiation or joint decision-making. Clearly, the experience of participation had been favourable.

A close examination of figures 68 and 69 reveals that this improvement in the preferences by managers for improved future levels of participation in the **planning** phase of technological change varies from one country to another. It is particularly evident in the case of the middle-ranking countries, especially **Belgium**, the **United Kingdom, France, Spain, Greece** and **Italy**. Only in **Portugal** do we find that the improvement in future preferences of managers is small.

In the **implementation** phase, a similar pattern is evident, particularly in the cases of **France, Spain, Greece** and **Italy.** Although there is small improvement in the case of **Portugal,** the preference for increased levels of future participation is at best marginal. Another point to note is that **United Kingdom** managers show a substantial desire for consultation in the future; this reflects the move towards inculcating attitudes of identification with the aims of the enterprise, which we referred to in chapter VIII.

When we look at the preferences for the future of employee representatives in the **planning** phase, **all countries** with the exception of **Portugal** register a substantial shift in opinion. Another striking feature is that employee representatives in two particular countries - **Denmark** and **Germany** - show a marked preference for major decisions on technological change to be taken either by negotiation or by joint decision-making.

A very similar pattern exists for employee representatives in the **implementation** phase. All countries - again with the exception of **Portugal** - want more negotiation or joint decision-making and the numbers who envisage no involvement shrinks substantially.

Secondly, the figures reveal that whilst both managers and employee representatives throughout the European Community are generally in favour of improved levels of participation in the future, **there are differences of opinion between the two sides.** This is true in all countries and there is nothing surprising in this. The interests of management and employees are very different and we would expect

SUMMARY OF INDIVIDUAL COUNTRY RESULTS

Figure 70: Past participation in implementing for new technology in the EC countries - Managers

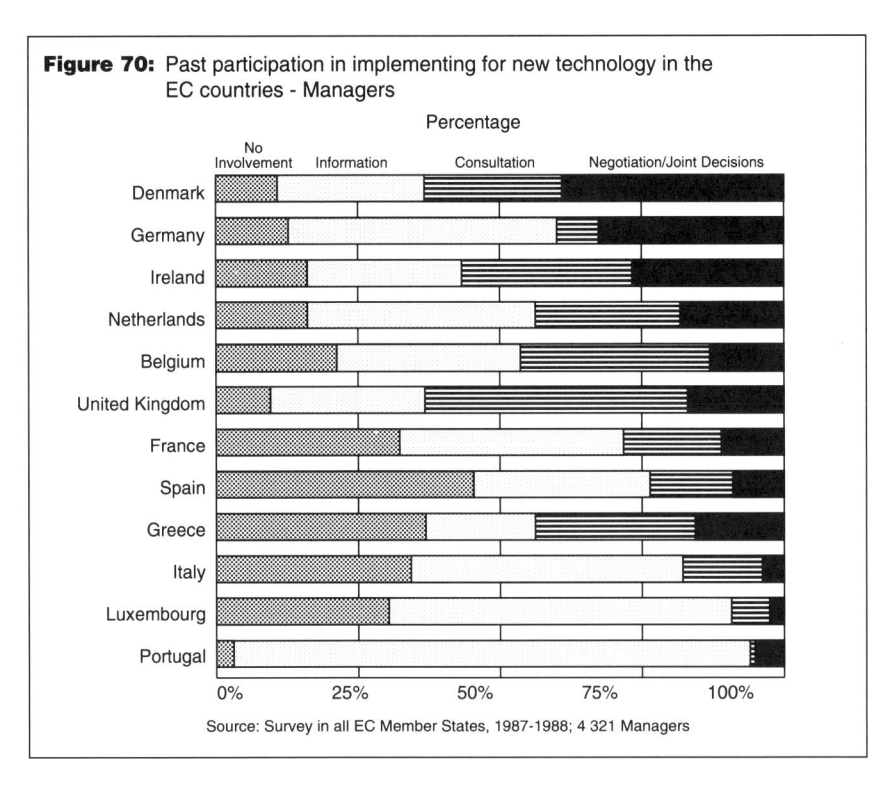

Percentage

No Involvement / Information / Consultation / Negotiation/Joint Decisions

Denmark, Germany, Ireland, Netherlands, Belgium, United Kingdom, France, Spain, Greece, Italy, Luxembourg, Portugal

Source: Survey in all EC Member States, 1987-1988; 4 321 Managers

Figure 71: Future Participation in implementing new technology in the EC countries - Managers

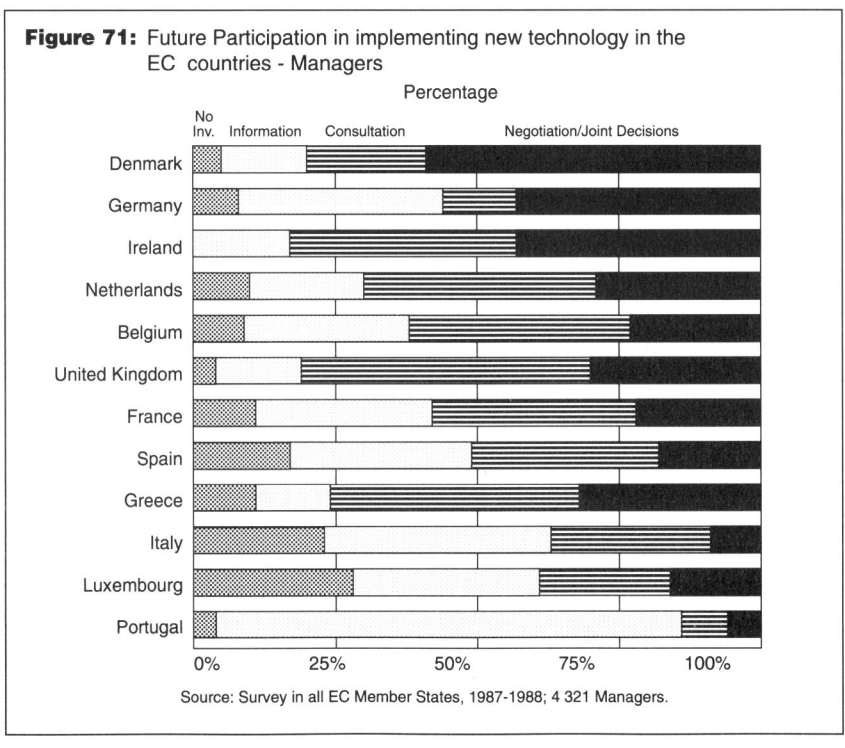

Percentage

No Inv. / Information / Consultation / Negotiation/Joint Decisions

Denmark, Germany, Ireland, Netherlands, Belgium, United Kingdom, France, Spain, Greece, Italy, Luxembourg, Portugal

Source: Survey in all EC Member States, 1987-1988; 4 321 Managers.

employees to seek greater influence over technological change than management is prepared to concede. However, such dissent between the two sides could herald conflict for the future. Even where the difference between the two sides is small this in itself should not be taken to mean that the potential for future conflict over issues presented by new technology is minimal; on the contrary, such apparent consensus might only conceal wider issues of disagreement when new technology is introduced into enterprises in the future.

In which countries is there the greatest lack of unanimity about future levels of participation between managers and employee representatives? **Denmark, Germany, Italy** and **Greece** are perhaps the most obvious countries where disagreement seems to be the strongest. In **Denmark** employee representatives have been used to negotiating issues thrown up by technological change for a considerable period of time aided by the centralised industrial relations system and the willingness of employers to negotiate on a wide range of matters through Co-operation Committees at a number of levels. They appear to believe that the scope of negotiation can still be expanded even though Danish managers are much more reluctant to move forward in this direction. At the same time, the Danish industrial relations system has proved to be remarkably conflict-free and has had a high degree of stability over several decades. It is therefore unlikely that this disagreement between the two sides of industry will lead to increased conflict in the future.

In **Germany** there is also a high measure of disagreement between both sides. As we pointed out in Chapter VIII, the German system of industrial relations has undergone some change during the 1980s. The DGB has been in the vanguard of the ETUC strategy of reducing working time and has treated the issue of technological change perhaps more seriously than any other trade union movement in the European Community. It is critical of what it sees as the limiting role of the German co-determination system and the absence of a trade union influence at shop-floor level. The German employers, for their part, are very much opposed to any improvement in the provisions for participation. Yet, as with **Denmark,** the German system of industrial relations has proved to be remarkably conflict-free and stable over the years and it would be surprising if the amount of disagreement between the two sides as to the future of participation should result in increased conflict.

It might be argued that **Italy** is a different case altogether. Here over three out of five managers envisage either no involvement at all or at best the mere provision of information on technological issues. Italian employee representatives have much greater aspirations; over a third of them want the **planning** phase of technological change to be decided by either

Summary of Individual Country Results

Figure 72: Past participation in planning for new technology in the EC countries - Employee representatives

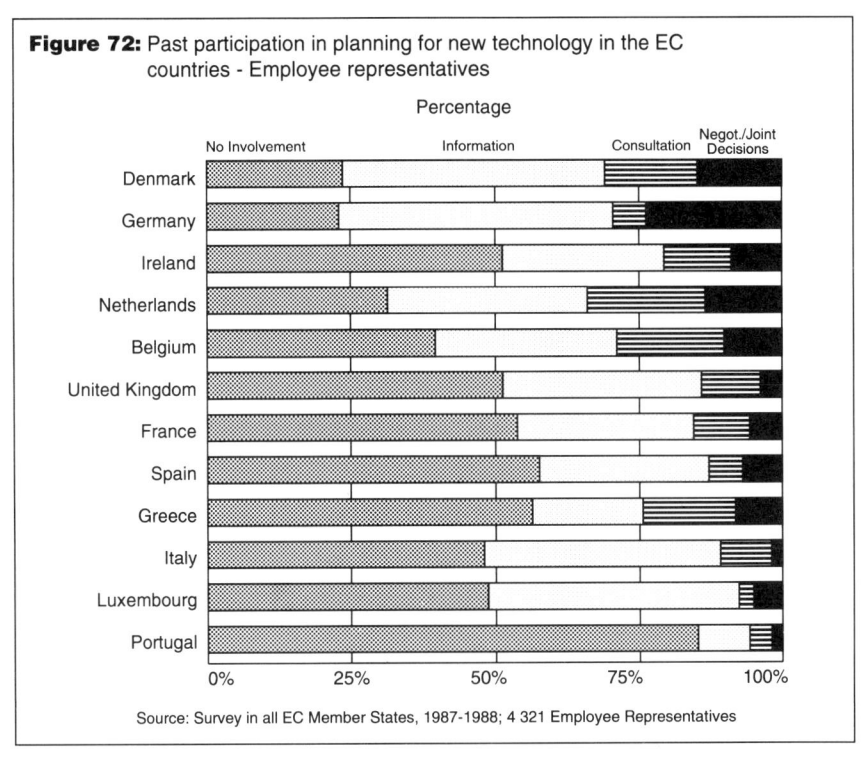

Source: Survey in all EC Member States, 1987-1988; 4 321 Employee Representatives

Figure 73: Future participation in planning for new technology in the EC countries - Employee representatives

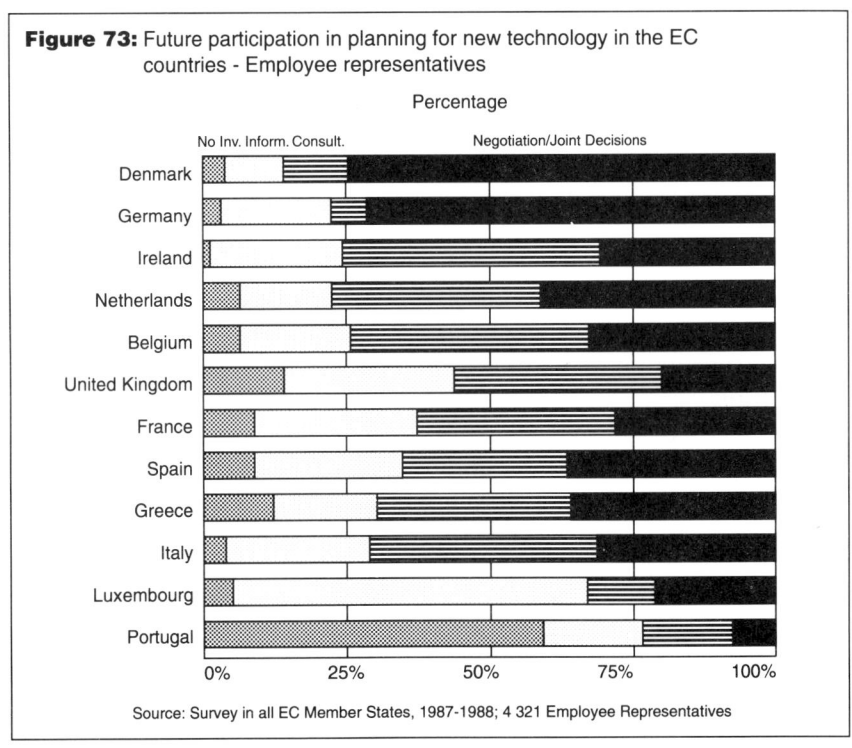

Source: Survey in all EC Member States, 1987-1988; 4 321 Employee Representatives

negotiation or co-determination, and just over a quarter of them are prepared to accept either no involvement or the mere receipt of information. Given the historical polarisation between the two sides of Italian industry and the "class" traditions of Italian trade unionism, this does not augur well for the future. On the other hand, there are signs of change in Italian industry, and there is evidence that a more flexible, co-operative industrial climate is emerging, slow though that process is.

In **Greece** the legacy of military dictatorship is still evident. The unions are inexperienced and poorly resourced; they have almost a blind faith in relying on favourable legislation from a sympathetic government. There are other factors here that also have to be considered: the large number of tiny enterprises; the sizeable "black" economy; the low level of technological diffusion and a polarisation in Greek society which is reflected in the workplace. It is still too early to make any firm predictions about what the effects will be of this difference in opinion between the two sides as to future participation levels.

5. The importance of statutory rights and legal regulations

It goes without saying that statutory rights and legal regulations are but one of the factors which are influential in shaping the opportunities for employee involvement in technological change.[1] How important are they? Our survey results can only give a limited pointer to answering this question. As we emphasised earlier, all our explanatory factors which are influential in shaping the opportunities for increased involvement by employees in technological change can apply at all levels; national level, sectoral level and at the workplace. Moreover, it is notoriously difficult to assess the impact of one factor to the exclusion of others.

Our survey results show that certain countries can have a very high degree of participation without necessarily having binding legislative provisions; **Denmark** is a good example. It can also be argued that **Ireland** and the **United Kingdom** have reasonable levels of participation. In Ireland and in the U.K. there is a voluntaristic tradition and the former country - given the limitations of the small number of interviews conducted - features in the group of countries in the upper part of our participation tables and has no legislation (at least covering those enterprises which featured in our interviewing sample) providing for participation in technological change. In the U.K. there is a striking willingness by British mangers to envisage a high degree of consultation on technological change in the future.

On the other hand, we have the example of **Italy**, where participation is low and where there is little in the way of legal regulation. Against this

Summary of Individual Country Results

Figure 74: Past participation in technology implementation in the EC countries - Employee representatives

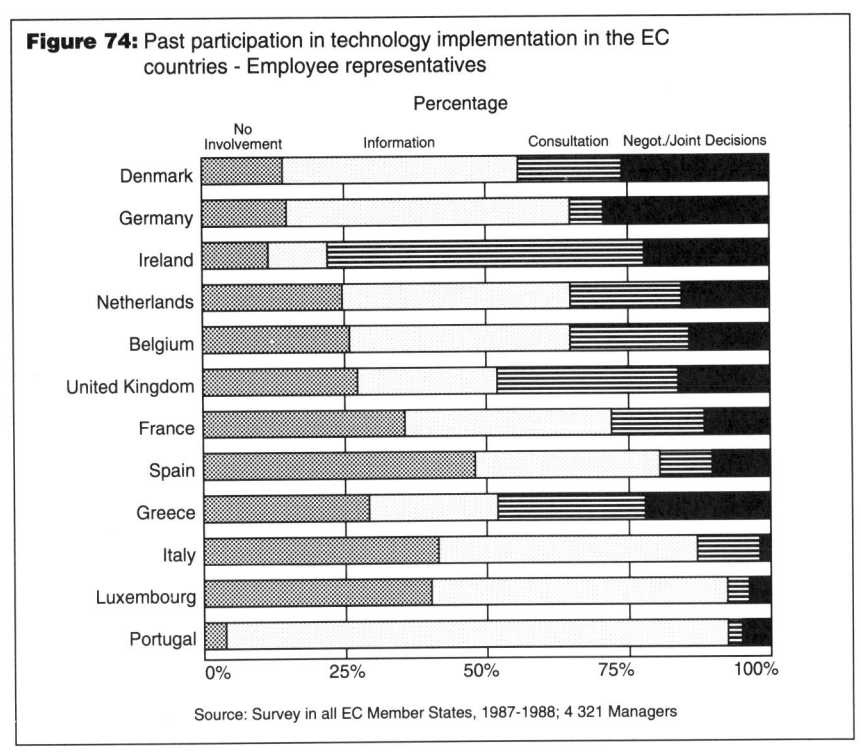

Source: Survey in all EC Member States, 1987-1988; 4 321 Managers

Figure 75: Future participation in implementing new technology in the EC countries - Employee representatives

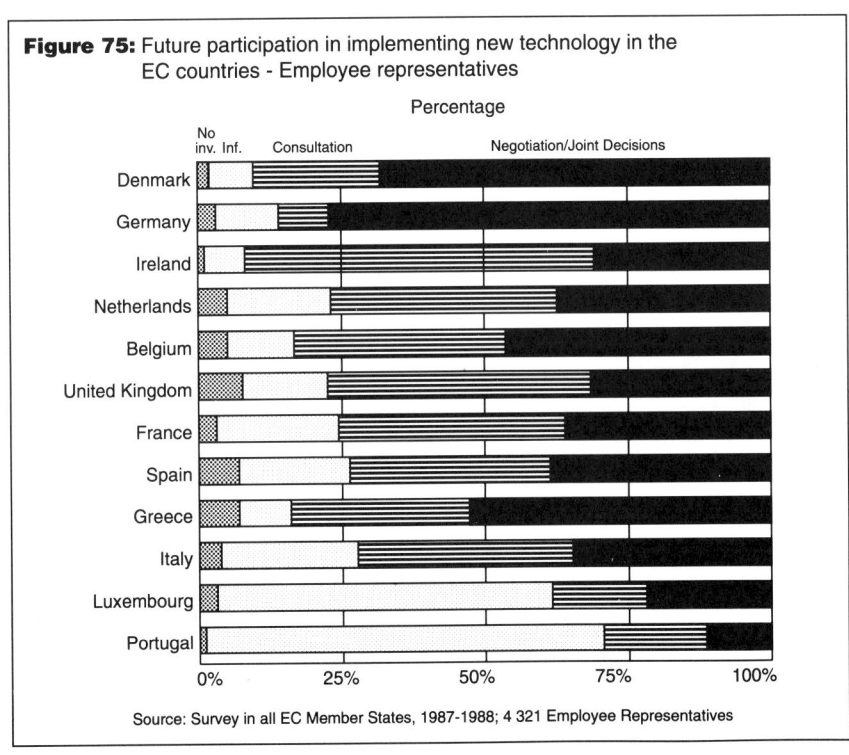

Source: Survey in all EC Member States, 1987-1988; 4 321 Employee Representatives

we also have to note that there are so many other factors in existence in this country which make it extremely difficult to arrive at any firm conclusions.

Underlying any attempt at evaluating the impact of legislation on levels of participation is the fact that, in many of the countries in our survey the legislation itself was limited in its provisions and as a result was ineffective. Several examples of this could be cited: the **Netherlands, Belgium, Luxembourg,** and **Spain.** In addition, we could cite the case of **Portugal**, where the law is so precise and detailed that trade unions, the form and conduct of collective bargaining and the operation of works councils are tightly regulated to the extent of complete inflexibility. However, in countries where the legacy of a dictatorial political régime still prevails, where the industrial relations system is still very much in its embryonic stage and where technological diffusion is low, legislation can have a very important impact on improving the levels of participation simply and quickly. **Greece, Portugal** and **Spain** are all good examples of this. In conclusion then, it is difficult to arrive at any firm conclusions as to the importance of statutory rights and legal regulation in shaping the opportunities for employee involvement in technological change for all the reasons cited above.

6. Assessing the results in the light of the Val Duchesse Joint Opinion of March 1987

The dialogue between the two sides of industry at European level (The European Trade Union Confederation [ETUC] on the one hand and the Union of Industrial and Employers' Confederations of Europe [UNICE] and the European Centre for Public Enterprises [CEEP] on the other) was revived by Jacques Delors at the beginning of 1985. It is known as the Val Duchesse social dialogue. This type of dialogue is covered by the Single European Act, Article 118B of which states that "the Commission shall endeavour to develop the dialogue between management and labour at European level which could, if the two sides consider it desirable, lead to relations based on agreement". The meeting on 12 November 1985 led to the setting up of two working parties, one of which looks at macroeconomic questions, while the other deals with microeconomic aspects. On 6 March 1987, the working party on microeconomic questions issued a joint opinion on the new technologies, dealing with training and motivation, information and consultation.

It is therefore important to assess these results in the light of the policies of the two major social partners at European level, UNICE and the ETUC. Whilst both social partners have very divergent views as to how participation in technological change should be conducted, they were both signatories to the Val Duchesse Social Dialogue Joint Opinion of the working party on 'Social dialogue and new technologies' concerning training and motivation, and information and consultation. The relevant clause of this agreement, against which our results should be assessed, reads as follows:

"Both sides take the view that, when technological changes which imply major consequences for the workforce are introduced into a firm, workers and/or their representatives should be informed and consulted in accordance with the laws, agreements and practices in force in the Community countries. This information must be timely.

In this context:

(a) information means the action of providing the workers and/or their representatives, at the level concerned, with relevant details of such changes, so as to enlighten them as to the firm's choices and the implications for the firm's workforce;

(b) consultation of the workers and/or their representatives, at the level concerned, means the action of gathering opinions and possible suggestions concerning the implications of such changes for the firm's workforce, more particularly as regards the effects on their employment and their working conditions."

We have interpreted this joint opinion to mean that employee representatives must be, **at the minimum, consulted** about new technology in **both** the planning and implementation phase of its introduction into enterprises. The mere provision of information about the introduction of new technology does **not**, therefore, fall within the guidelines laid down in the joint opinion. Of course, levels of participation involving negotiation or joint decision-making are superior to the Val Duchesse provisions.

How do all our countries rank in terms of the Val Duchesse guidelines? We first consider the situation that prevailed at the time when our interviewing took place in 1987-88. If we consider the views of managers we find that in the **planning** phase only in **Denmark** and **Ireland** did half of the managers concerned report that consultation, or stronger levels of involvement, occurred. A middle-ranking group of countries which included **Germany**, the **Netherlands, Belgium,** the **United Kingdom, France, Spain** and **Greece** all had no more than 30% of managers who claimed levels of participation which were within the Val Duchesse guidelines. There was almost no compliance with Val Duchesse in the cases of **Italy, Luxembourg** and **Portugal.**

When we look at the responses of employee representatives the compliance with the Val Duchesse provisions in the **planning** phase looks even bleaker. Only in the case of the **Netherlands** do more than a third of employee representatives report that Val Duchesse is being observed. There is a middle-ranking group of countries including **Denmark, Belgium, Germany** and **Greece** where about a quarter of respondents state that consultation and higher levels pertain. All the other countries fall in the range of ten per cent and below.

As we would expect, the compliance with Val Duchesse is better in the **implementation** stage. Yet here only in **Denmark, Ireland** and the **U.K.** do we find a **majority** of managers who state that the **implementation** stage of technological change is characterised by at least consultation or higher levels of involvement. There is a middle-ranking group of countries where the percentage of managers reporting Val Duchesse compliance clusters around 40%. Below that the percentage range decreases down to just above 5% (in the case of **Portugal**).

Employee representatives again are more sceptical compared to their management counterparts except in the case of **Ireland.** Here it is notable that over 75% (!) of employee representatives in **Ireland** state that consultation and higher levels are evident. However, this is the only country where we find that managers' opinions are not contradicted by employee representatives. In all other cases there is **not one single country where a majority of employee representatives state that Val Duchesse provisions for the implementation phase apply.**

One question that might legitimately be asked is "whom are we to believe?": managers or employee representatives? If we take a more sceptical view and accept the views of the workforce representatives as a more accurate statement of reality, then the present situation with regard to the observance of Val Duchesse guidelines on employee participation is bleak indeed.

Is there any likelihood that the Val Duchesse guidelines will be observed in the future? Three immediate conclusions can be derived from our data for the **planning** stage of technological change, given their past experience. Firstly, the most striking conclusion that can be made is that there is a significant shift of opinion among both sides in favour of a greater level of participation in all the countries surveyed. It is also evident that employee representatives would prefer a higher level of participation than their management counterparts; this is only to be expected. Thirdly, the pattern of preferences for the future, in terms of the ranking of countries, bears a close resemblance to that of the past.

The most useful way of assessing the likelihood of compliance with Val Duchesse guidelines in the future is to consider the future preferences of managers. It is their attitudes, after all, which are important in determining the probable levels of participation that might emerge in the future. There are five countries from our sample in the **planning** phase where over half of managers indicated that Val Duchesse guidelines are most likely to be observed: **Denmark, Ireland,** the **Netherlands** the **United Kingdom** and **Greece**. A striking observation that can be made is that **Germany is not included in this list of more highly-ranked countries**; only two out of five managers in this country anticipate that Val Duchesse guidelines will operate in the future. One can only explain this by referring to the strong opposition that German managers have shown throughout the 1980s to any improvement in works councils legislation in favour of employees. There is a middle-ranking group of countries where the percentages of managers indicating likely compliance with Val Duchesse are around the 30-40 range: **France, Spain, Germany, Luxembourg** and **Italy**. In **Portugal** the percentage of managers was just below 25%.

In the **implementation** phase of technological change the position with respect to Val Duchesse guidelines is much healthier. Here there is a similar pattern to that of the planning stage insofar as the future preferences of managers are concerned. **Denmark** and **Germany** show the highest preference for strong forms of participation. A **majority** of managers in all countries except **Italy**, **Luxembourg** and **Portugal** envisaged levels of participation in the future which consisted of participation levels involving consultation or stronger forms. On the negative side, the levels of no involvement are still worrying. **France, Greece, Italy, Luxembourg** and **Spain** all have results which show that over 10% of managers intend to introduce new technology in the future into their enterprises unilaterally without any employee involvement whatsoever. **Portugal** again appears to produce some surprising results; whilst just over a quarter of managers there seem prepared to observe Val Duchesse guidelines in the implementation phase, the level of no involvement is the second lowest of all the European Community countries. Nevertheless, the great majority of Portugese managers appear to believe in restricting participation simply to the supply of information.

Any conclusions that can be made about the likelihood of Val Duchess social dialogue guidelines being observed in European Community countries in the future must inevitably be couched in uncertainty. It is certainly true to say that managers are much more likely to support Val Duchesse when major decisions about the introduction of new technology have already been made and when it is being installed in the workplace. It is then that managers need to consult with their workforce to make the fullest use of the investment that has been made. At the same time, it should be said that Val Duchesse guidelines are not just confined to the implementation stage of the introduction of new technology; they are meant to apply to the planning stage as well.

Those countries where Val Duchesse guidelines are more likely to be observed in the future all appear to have more centralised industrial relations systems, a more co-operative management style, long-standing provisions on works councils, and a trade union movement which is not weakened by divisions and inter-union rivalry or is reluctant to embark on participation for fear of identifying itself with management's objectives. Our results can only give a broad indication as to the contours of participation levels in the future, and it would be unwise to arrive at any definitive conclusions at a time when industrial relations throughout the European Community is undergoing fundamental change.

7. The policies of the ETUC

As we mentioned in earlier chapters, the European Trade Union Confederation has policies on participation in technological change which are far in excess of the Val Duchesse joint opinion. For example, at its Sixth Congress in Stockholm in 1988 it demanded "the right of employee representatives to be fully informed, consulted and also to negotiate on **all** important company matters before decisions are taken ... equal participation by employee representatives in all company decisions of significance to the workforce...extension of decision-making rights at all levels of decision-making according to the organisation of companies ... the employee representatives at all plants must accordingly ... have the right to be informed and consulted on company planning, to **negotiate** and to represent their interests **jointly** at European level".

The ETUC had previously stated at its Fifth Congress in Milan in 1985 that "effective negotiation on technological change often means **negotiating with management before** the introduction has been planned. It means re-appraising the knowledge and requirements of working parties in the production process and calling in outside experts to demystify the new technologies". We have therefore interpreted the ETUC position to mean that technological change must be characterised by negotiation and/or joint decision-making in both phases of the introduction of new technology.

The best way of assessing the support of employee representatives for the ETUC position on participation in technological change is to measure the preferences of workplace representatives in our survey for the kind of participation they seek in the future. What percentage of them in each country opt in their preferences for negotiation or co-determination in both the planning and implementation stages? **Denmark** and **Germany** are clearly well ahead of every other country as far as the **planning** phase of new technology is concerned. Nearly three out of every five employee representatives in these two countries would like major decisions on the choice, design and the planning of technological innovation to be made either by negotiation with management or by co-determination. In each case there is a stronger preference for co-determination, but more so in Germany - reflecting the importance of the German co-determination system. There is a middle-ranking group of countries where the preference for the ETUC position hovers around the 30-40 per cent mark - **Ireland**, the **Netherlands, Belgium, France, Spain, Greece** and **Italy**. Employee representatives in the **United Kingdom, Luxembourg** and **Portugal** all have less than one in four employee representatives apparently subscribing to the ETUC position.

How can this variation in ranking be accounted for? Both **Denmark** and **Germany** have trade union movements which have been relatively untouched by the decline in membership that has beset trade unions throughout the EC during the 1980s. Whilst the density of unionism in Denmark far surpasses that of Germany, the DGB in Germany has been in the vanguard of spearheading and publicising ETUC and trade union policies generally, and there is consequently a high awareness of the issues posed by new technology in the minds of German workplace representatives (most of whom are union members). In all the other countries, trade union membership has been declining recently as a result of many of the changes in industrial relations throughout the European Community, and many of them are either weak or divided. It is clear that there is a **long way to go before the ETUC can legitimately claim to have the support of the great majority of its rank-and-file representatives** at the level of European enterprises.

In the **implementation** phase the picture is slightly improved. Here, the ranking of the countries almost mirrors exactly that of the planning phase. Only in **Denmark, Germany** and **Greece** do a **majority** of employee representatives appear to support the ETUC position. In relation to the Val Duchesse guidelines however, the **majority** of employee representatives in all countries except **Luxembourg** and **Portugal** all want these guidelines to be observed in the future in both phases of the introduction of new technology into enterprises.

Are we to conclude that the support for the policies of the ETUC among workplace representatives leaves much to be desired? It is certainly concentrated in two countries, and in all the other countries it is confined to a minority of workplace representatives. We have to be treat the survey results with some caution for two reasons. First, this report does not differentiate between union members and non-unionists. If we had done so, we might have found that the ETUC position gained much stronger support among trade union members than non-unionists. Secondly, in our interviewing, employee representatives were **not** reminded of what ETUC policy on technological change was before being asked a different question such as: "Do you support the ETUC policies of negotiation and/or co-determination in all aspects of technological change?". If this question had been asked we might have received different responses.

8. The policies of UNICE

UNICE's policy on participation in technological change is linked to its general policy on the Social Dimension of the internal market in the Community. In its own words: "Priority must be given to improving the level and quality of employment. All measures likely to increase the burden of social charges on European businesses or to reduce their managerial flexibility have to be avoided, since this would seriously threaten their position relative to their main competitors inside and outside the Community ..."

UNICE encourages experimentation in the field of industrial relations, since it believes that nobody yet knows what the best methods will be in the world of new technologies and of the Internal Market into the next century. It views the Social Dialogue as an important forum in which the two sides of industry and the Commission together review developments and steps still to be taken to realise the Internal Market, and discuss social problems in their economic context. UNICE is vigorously opposed to any notion of standardising participation and industrial relations practices throughout the Community but is quite happy to allow participation practices to take place (including negotiation and co-determination) providing they satisfy the needs of both sides of industry according to local conditions, traditions and legislation in particular countries. In short, they strongly oppose the imposition of any particular form of participation on all Member States by means of EC-level legislation. They therefore view the Social Dialogue joint opinion being used as a basis for Community legislation without the consent of all the parties concerned.

9. Conclusions

Perhaps the most important conclusion that emerges from our survey results in individual countries of the European Community is the sheer diversity in the forms of participation in new technology that takes place. We sought to explain the reasons for this diversity by the development of an explanatory framework which has proved to be very useful in analysing the great differences that exist from country to country. Whilst this explanatory framework has been of value, it cannot claim to explain every variation down to the finest detail. The factors will operate to different degrees at national, sectoral and enterprise levels in each country, and it is a momentous task to adequately explain the vast diversity in participation that exists across the Community. Customs and traditions vary and the perceptions of the two sides of industry about the nature and meaning of participation will be different in country to country either because of subtleties of language or because their perceptions about particular types of participation practices have been shaped by factors other than their own experience.

What can also be concluded is that taking all our explanatory factors together, it is clear that the types of participation that exist in Europe, and which will emerge in the future, are a product of the way the contours of each Member State's industrial relations system have been shaped by wider political, economic, social and historical forces.

Another important conclusion is that there is evidence that all countries of the Community are moving in different degrees towards a greater level of participation in the future. While the pace of change might not necessarily be to the satisfaction of everyone, our survey shows that the experience of participation by both sides engenders favourable attitudes towards more intense forms of participation in new technology. It is our hope that this trend towards greater participation levels in the future which has been highlighted from our survey results will prove to be true.

NOTES

1. Legal Regulations on Participation and the Intensity of Participation in Europe, Michael Gold and Mark Hall, European Foundation for the Improvement of Living and Working Conditions, Dublin, 1990

Technical Appendix

The Introduction of New Technology Survey

Prepared for: The European Foundation

Prepared by: The Harris Research Centre
Holbrooke House,
Holbrooke Place,
34-38 Hill Rise,
Richmond,
Surrey TW10 6UA

Telephone: 01 948 5011

Fax: 01 948 6335

Telex: 24403

Technical Appendix

Contents

1. Survey Scope...238
2. Sampling...239
   - 2.1 Survey Universe...............................239
   - 2.2 Sample Selection..............................241
   - 2.3 Interview Procedure..........................244
3. Response Rates....................................245
   - 3.1 Response Categories........................245
   - 3.2 Response Rate Analysis....................262
   - 3.3 Commentary on Response..................264
4. Fieldwork Dates...................................265
5. Weighting Procedures..........................265
6. Agencies Involved................................285
7. Questionnaire Translations...................285
8. Levels of Technology by Sector..............286

1. Survey Scope

The survey covered in all 12 countries of the EC. It was carried out in two waves, the first wave covering 5 countries in 1987 and the second wave covering the remaining 7 countries in 1988.

For wave 1, a sample size of 1000 interviews (500 pairs) was established for each country, split as follows:

Retailing	120 (60 pairs)
Banking/Insurance	240 (120 pairs)
Mechanical Engineering	520 (260 pairs)
Electronics	120 (60 pairs)

However, in Denmark it proved impossible to meet the target samples because of a lack of establishments eligible in terms of size and in total only 628 interviews were finally completed there.

For wave 2, therefore, sample sizes were set for each country based as far as possible on establishment structure data. However, size data at a sector level of detail is limited in nearly all countries and despite attempts to be realistic, it still proved impossible to complete the targets in four countries, even after the size parameters were reduced. Ireland was by far the worst hit, since the country proved to have very few establishments which met the size or other criteria set for the survey.

The table below summarises the targets set and the achievement rates for wave 2. A more detailed analysis of results is given in section 3.

	TARGET	ACHIEVED	%
Belgium	700	704	100
Greece	400	254	64
Ireland	400	76	19
Luxembourg	200	120	60
Netherlands	750	518	69
Portugal	400	400	100
Spain	600	606	100

The parameters for interviewing in all countries were established as follows:

— All interviews had to be paired; for each management representative there had to be a complementary employee representative interview.

— A maximum of two pairs of interviews could be carried out in each establishment. However, if more than one pair of interviews was done, each pair had to relate to different technologies and the interview respondents both had to be different people.

— Within one company, no more than four establishments could be interviewed.

2. Sampling

2.1 Survey Universe

The survey universe was defined as those establishments meeting specified criteria in each country.

The specified criteria were on three levels:

A. By sector
B. By new technology introduction/involvement of employee representative within the establishment
C. By size of establishment.

The criteria included in points A and B remained constant across all countries in both waves; those in point C varied by sector and by country, as shown below.

A. Sectors

The following sectors defined in all countries by NACE codes, formed the basis of the survey:

	NACE Codes
SERVICE SECTORS	
Retailing	641-649
Finance, Banking, Insurance	811-813
	821-823
	832
SERVICE SECTORS	
Mechanical Engineering	322-328
	314
Electronics	330
	342-347
	371-372

B. New Technology/Employee representation criteria

All companies interviewed in the survey had to meet all the following criteria. Failure on any of them rendered the company ineligible for inclusion:

— Must have introduced at least one new technology from a specific list of technologies within the last five years.

— Must have a formal system of employee representation

— Must have involved their employee representative(s) in the introduction of new technology

C. Size of establishment

The main determinant of establishment size was the number of employees working there. The minimum number of employees necessary for inclusion varied according to sector and/or according to the type of establishment interviewed. The following are the minimum criteria established for Wave 1 countries:

SECTOR	ESTABLISHMENT TYPE	MIN. NO. OF EMPS.
Retailing	Head Office	100
	Supermarket/sales area of 200 sq m +	
Banking	Head Office	100
	Branch	50
Insurance	Head Office	100
Mechanical Engineering	Any	100
Electronics	Any	50

In Wave 2 initially, the same size criteria were adopted. However, in these mainly small countries, the size criteria proved to be over-ambitious and the following minimum size criteria variations wre finally adopted (although the companies concerned remained subject to meeting the criteria outlined in points A and B above).

TECHNICAL APPENDIX

	MIN. ESTAB. EMP. SIZE
BELGIUM	
Mechanical Engineering	40
Electronics	25
GREECE	
Banking and Insurance	15
Mechanical Engineering	40
Electronics	25
IRELAND	
Retail	15
Banking and Finance	15
Mechanical Engineering	40
Electronics	25
LUXEMBOURG	
Banking	15
Mechanical Engineering	40
Electronics	25
NETHERLANDS	
No changes	
PORTUGAL	
Mechanical Engineering	40
Electronics	25
SPAIN	
Insurance	15
Mechanical Engineering	40
Electronics	25

2.2 Sample selection

In each country, the sample sources used to construct initial address lists are shown on Tables A and B on the following pages. For those countries where reliable official establishment sources were available, no other source was used. For most other countries, agencies used the most comprehensive source available, cross-referencing a 'core' listing where possible with the most recent trade listings or other data sources, to ensure a complete starting address list.

Address lists were then stratified by size with NACE codes and initial samples drawn systematically with equal probability of selection. If the initial sample was exhausted without quota completion, the process was repeated on the remaining sample. However, in many countries, particularly the smaller ones in wave 2, the number of eligible establishments proved very limited in one or more sectors, either in terms of size or in terms of the other eligibility criteria. In these countries/sectors, the total universe was in fact contacted (see Response Rates, section 3 for details).

TABLE A—SAMPLING SOURCES—WAVE 1

	UK	FRANCE	DENMARK	ITALY	GERMANY
RETAIL	Retail Directory 1986. Supermarkets & Superstores. Food Trade Yearbook. Yellow Pages.	INSEE	Press: Danske Supermarkeder 1986.	Caratteri Strutturali Del Sistema Distributivo in Italia. Some Editorials of "Largo Consumo".	Glending & Lehning Top 200
BANKING	Building Society Yearbook. World Insurance 1986. Insurance Almanack.	INSEE	Membership Lists	Annuario Dello Aziende di Credito e Finanziarie by ABI 1986.	Deutsche Bundesbank. Verzeichnis der Kredit-institute.
INSURANCE	The Banker's Almanack. The Banker. Times 1000. Yellow Pages	INSEE	Forskirings-handbogen	Annuario Italiani delle Imprese Assicuratrici by A.N.I.A. 1986.	Hoppenstedt: Versicherungs-Jahrbuch
ELECTRONICS	Industrial Market Locations	INSEE	Erhvervs-registeret at Danmarks Statistik	Kompass. Directory of ANIMA Member Firms.	Hoppenstedt: Deutsche Grossunternehmen. Kompass.
ELECTRONICS	Industrial Market Locations	INSEE		Kompass. Directory of ANIE Member Firms	Elektro-Einskaufsführer

TABLE B—SAMPLING SOURCES—WAVE 2

ALL SECTORS	
Belgium	Social Security Database
Greece	ICAP—Dun & Bradstreet (except Retailing: Nielsen Census of Retail Trade)
Ireland	Dun & Bradstreet Top 500 companies listing.
Luxembourg	Repertoire des Entreprises Luxembourgeoises
Netherlands	Chamber of Commerce database.
Portugal	Dun & Bradstreet
Spain	Establishments Census 1980, from the Institute Nacional Estadistica.

2.3 Interview procedure

In the first wave, the initial contact with the address was made by a telephone interview with the personnel manager, to determine establishment eligibility and to find out the names of suitable respondents involved with the introduction of new technology. With his permission, a personalised letter and self-completion company questionnaire were then sent to the nominated respondents followed by a call approximately one week later to make an appointment for the main stage interview.

However, this procedure proved somewhat unwieldly and in some cases the effect of this initial approach by letter to the nominated respondent was to increase the likelihood of a refusal. For wave 2 therefore the first contact with nominated respondents was also made by phone, and once agreement to participate had been established the introductory letter and questionnaire were then sent either by post or fax and a follow-up call made soon after to continue the procedure and make firm appointments.

The personnel manager in each establishment was thus used as the first point of contact and asked to nominate suitable pairs of respondents involved in the introduction of new technology. Where more than two of the specified new technologies had been introduced, the personnel manager was asked to select the two which had had the greatest impact and with which different management and employee representatives had been involved. In all cases, it was made clear that it was not technical experts which were required for interview, but those both from the personnel/operational aspects of the introduction.

Whilst the above describes the overall approach taken for the survey in simple terms, in many cases the contact stages were very much more complex, and a flexible approach was necessary, depending on the attitude of the personnel manager and subsequently on the attitudes of nominated respondents. In all countries, the fieldwork agencies made very possible effort to secure pairs of interviews, if necessary contacting Head Office establishments for permission clearance, sending personalised letters or talking to unions, and re-contacting potential respondents after cancelled appointments. All countries experienced some difficulties during the contact stages of the survey, although they varied in nature and severity, and all took appropriate actions.

Once appointments had been made, all full interviews were carried out face-to-face by executive interviewers, that is by interviewers experienced in business interviewing. Most were carried out in the workplace, although when requested a small minority were conducted with employee representatives at home.

3. Response Rates

3.1 Response categories

The response rates in terms of absolute numbers, both overall and by country, are given on the tables C1-C14. Once non-eligible and non-used samples are discounted from the total addresses issued then the response rate in terms of the relationship between achievement and refusals amongst eligible companies can be considered satisfactory given the difficulties inherent in successfully achieving paired interviews. (see 3.2 below and Table D for a calculation of the achievement rate amongst eligible establishments).

However, before examining the response rates tables and the response rate calculations p. 248-261 is important to bear in mind the following points in respect of the response categories shown on the tables.

(a) NOT-USED SAMPLE: This category includes all establishments remaining after quotas had been completed, including the surplus establishments of multi-branch companies which remained unused after the maximum permissible number of branches per company had been interviewed. In calculating the response rate amongst eligible companies, these establishments have been deducted from the total issued sample as non-eligible sample.

(b) NON-CONTACTS: Non-contacts were of two types. The first line on the table shows the number of addresses which turned out to be totally unusable, including establishments which had closed down, gone out of business, the number was unobtainable etc. The second line shows companies where no contact with the appropriate personnel manager was ever made, despite up to 10 attempts. These occurred mainly in the major countries, and in a sense could be considered as non-used/over quota addresses, since they represent the balance left over after targets were completed.

In calculating response rates below, all non-contacts have been counted as non-used/non-eligible sample and deducted from the initial addresses issued.

(c) NON-ELIGIBLE SAMPLE: This category includes all establishments identified as non-eligible during the screening process, and a breakdown of the different categories of non-eligibility is also given. It will be noted, particularly in Wave 2 countries, that the size of the establishment (i.e. the number of employees) is a major component of non-eligibility. This reflects one of the difficulties of establishment sampling, in that the accuracy and currency of establishment information available from many sampling sources is limited, in part because of rapid fluctuations in employee size at establishment level particularly amongst smaller companies, and in part because most sampling sources are primarily accurate at Company, rather than establishment level. With few large companies or large companies or large company establishments, many of the data sources for the smaller countries are particularly limited by such problems, though even in the major countries, size inaccuracy constitute a major part of sample ineligibility.

(d) PRE-SCREENING REFUSALS: Pre-screening refusals constituted a major problem in nearly all markets. Some of these refusals, mainly in large companies, were of the 'company policy' type (i.e. it is company policy not to take part part in surveys of any kind). Because company policy refusals mainly come from large multi-establishment companies, such refusals have a disproportionate effect to the number of establishments they involve—one company policy refusal eliminates all establishments in that company. Certainly the company refusal of two of the clearing banks in the UK and two of the major supermarket chains in France significantly reduced the number of available establishments in those sectors in those countries.

However, some pre-screening refusals were of the 'too busy' type rather than 'company policy' type, although occasionally refusals were made on the grounds of subject sensitivity, an attitude which varied greatly by country and by sector.

Pre-screening refusals constitute a difficulty when calculating the eligible universe, because as they were not screened, such establishments cannot be assessed for eligibility. The calculation of eligible establishments shown below therefore includes a proportion of these establishments, assuming the same distribution of eligible/non-eligible establishments would occur amongst pre-screening refusal establishments as occurred amongst screened establishments.

(e) POST-SCREENING REFUSALS: This category contains all refusals from screened eligible establishments, broken down by the point at which the refusal occurred. It will be noted that many of these refusals in the smaller countries in wave 2 occurred when the interviewer actually turned up for an appointment, either because one side of the pair had had second thoughts on the grounds of subject sensitivity or alternatively simply because of time limitations. Cancelled appointments caused a lot of problems with the timing schedules and resulted in large numbers of single interviews being done but discarded. However, in the larger countries, the main point of refusals was during the screening process, a difference which probably reflects establishment size differences.

(d) NUMBER OF INTERVIEW POINTS/INTERVIEWS: The response rate tables finally show the number of different establishments interviewed during the survey (discounting any multiple interviews) plus the number of pairs interviews actually carried out during the survey (i.e. including multiple interviews). After data processing a very small number of pairs of interviews were discarded, because of incomplete or inconsistent data sets and the final usable sample of pairs of interviews is also shown.

TABLE C1—RESPONSE RATE ANALYSIS—ACTUAL FIGURES COUNTRY SUMMARY—WAVE 1

	DENMARK	FRANCE	GERMANY	ITALY	U.K.	TOTAL
Initial addresses	440	2369	4283	3381	1848	12321
Not used (over quota)	—	—	—	—	—	—
Non-contacts (no business, etc.)	—	145	170	659	78	1052
Non-contacts (respondent not available)	50	383	545	996	26	2000
Non-eligible						
—Total	70	500	2016	993	520	4099
—Sector	3	62	593	216	64	938
—Size	4	251	1113	386	162	1916
—No new technology	39	135	117	65	90	446
—No Emp. Rep.	4	52	142	183	146	527
—No involvement	20	—	51	143	58	272
Pre-screening refusals	—	352	499	118	372	1341
Post screening refusals	112	641	527	233	390	1903
—At/during screening	69	515	401	201	290	1476
—After letter received	31	94	90	15	72	302
—When interviewer arrived	12	32	36	17	28	125
Total estabs. interviewed	208	350	526	382	462	1928
Total interviews (pairs)	314	513	555	452	493	2327
Total pairs after D.P.	314	513	555	452	493	2327
Total letters sent	251	476	652	414	534	2727

For description of category inclusions, please see text.

TABLE C2—RESPONSE RATE ANALYSIS—ACTUAL FIGURES COUNTRY SUMMARY—WAVE 2

	BELGIUM	GREECE	IRELAND	LUX.	NETHS.	PORTUGAL	SPAIN	TOTAL
Initial addresses	1102	653	440	172	1705	1888	1504	7464
Not used (over quota)	148	29	—	—	—	344	—521	
Non-contacts (no business, etc.)	152	27	16	1	197	11	58	462
Non-contacts (respondent not available)	—	—	—	—	147	—	31	178
Non-eligible								
—Total	212	275	323	68	411	1327	689	3305
—Sector	31	5	61	7	103	122	151	480
—Size	122	243	109	47	186	521	514	1742
—No new technology	17	9	26	5	33	151	24	265
—No Emp. Rep.	19	16	67	9	55	374	—	540
—No involvement	23	2	60	—	34	159	—	278
Pre-screening refusals	264	13	18	38	39	—	91	743
Post screening refusals	65	196	45	16	402	88	388	1200
—At/during screening	5	—	—	—	260	25	254	544
—After letter received	24	29	—	6	98	19	73	249
—When interviewer arrived	36	167	45	10	44	44	61	407
Total estabs. interviewed	261	113	38	49	229	118	247	1055
Total interviews (pairs)	352	128	38	61	259	200	303	1336
Total pairs after D.P.	348	127	38	61	259	200	303	1336
Total letters sent	326	338	—	23	371	181	381	1620

For description of category inclusions, please see text.

TABLE C3—RESPONSE RATE ANALYSIS—ACTUAL FIGURES

DENMARK (Wave 1)	RETAIL	BANKS	INSUR-ANCE	MECH. ENG.	ELEC-TRONICS	TOTAL
Initial addresses	66	45	22	249	58	440
Not used (over quota)	—	—	—	—	—	—
Non-contacts (no business, etc.)	—	—	—	—	—	—
Non-contacts (respondent not available)	1	9	4	28	8	50
Non-eligible						
—Total	22	4	2	32	10	70
—Sector	1	—	—	—	2	3
—Size	1	1	—	2	—	4
—No new technology	9	3	—	21	6	39
—No Emp. Rep.	2	—	—	1	1	4
—No involvement	9	—	2	8	1	20
Pre-screening refusals	—	—	—	—	—	—
Post screening refusals	10	7	1	89	5	112
—At/during screening	3	4	1	58	3	69
—After letter received	4	3	—	22	2	31
—When interviewer arrived	3	—	—	9	—	12
Total estabs. interviewed	33	25	15	100	35	208
Total interviews (pairs)	48	41	27	152	46	314
Total pairs after D.P.	48	41	27	152	46	314
Total letters sent	40	28	15	131	37	251

For description of category inclusions, please see text.

Technical Appendix

TABLE C4—RESPONSE RATE ANALYSIS—ACTUAL FIGURES

FRANCE (Wave 1)	RETAIL	BANKS	INSUR-ANCE	MECH. ENG.	ELEC-TRONICS	TOTAL
Initial addresses	533	527		763	546	2369
Not used (over quota)	—	—		—	—	—
Non-contacts (no business, etc.)	34	28		32	51	145
Non-contacts (respondent not available)	36	101		143	103	383
Non-eligible						
—Total	135	91		142	131	500
—Sector	4	12		15	31	62
—Size	66	60		75	50	251
—No new technology	38	19		38	40	135
—No Emp. Rep.	28	—		14	10	52
—No involvement	—	—		—	—	—
Pre-screening refusals	101	72		88	91	352
Post screening refusals	147	138		279	77	641
—At/during screening	110	102		251	52	515
—After letter received	30	21		22	21	94
—When interviewer arrived	7	15		6	4	32
Total estabs. interviewed	79	59	40	79	93	350
Total interviews (pairs)	122	90	50	126	125	513
Total pairs after D.P.	122	90	50	126	125	513
Total letters sent	116	135		107	118	476

For description of category inclusions, please see text.

TABLE C5—RESPONSE RATE ANALYSIS—ACTUAL FIGURES

GERMANY (Wave 1)	RETAIL	BANKS	INSUR-ANCE	MECH. ENG.	ELEC-TRONICS	TOTAL
Initial addresses	213	191	271	2706	902	4283
Not used (over quota)	—	—	—	—	—	—
Non-contacts (no business, etc.)	3	1	6	121	439	170
Non-contacts (respondent not available)	5	6	13	458	63	545
Non-eligible						
—Total	92	46	88	1351	39	2016
—Sector	40	—	7	335	211	593
—Size	29	45	58	820	161	1113
—No new technology	2	—	16	78	21	117
—No Emp. Rep.	15	1	7	80	39	142
—No involvement	6	—	—	38	7	51
Pre-screening refusals	13	12	13	304	157	499
Post screening refusals	48	36	84	227	132	527
—At/during screening	25	14	62	201	99	401
—After letter received	18	19	13	17	23	90
—When interviewer arrived	5	3	9	9	10	36
Total estabs. interviewed	52	90	67	254	72	526
Total interviews (pairs)	53	90	73	264	75	555
Total pairs after D.P.	53	90	73	264	75	555
Total letters sent	75	112	89	271	105	652

For description of category inclusions, please see text.

TABLE C6—RESPONSE RATE ANALYSIS—ACTUAL FIGURES

ITALY (Wave 1)	RETAIL	BANKS	INSUR-ANCE	MECH. ENG.	ELEC-TRONICS	TOTAL
Initial addresses	418	530		1952	481	3381
Not used (over quota)	—	—		—	—	—
Non-contacts (no business, etc.)	111	128		315	105	659
Non-contacts (respondent not available)	67	41		693	195	996
Non-eligible						
—Total	113	226		542	112	993
—Sector	8	15		154	39	216
—Size	69	156		146	15	386
—No new technology	3	12		36	14	65
—No Emp. Rep.	20	22		121	20	183
—No involvement	13	21		85	24	143
Pre-screening refusals	23	16		60	19	118
Post screening refusals	62	37		78	56	233
—At/during screening	55	30		66	50	201
—After letter received	4	3		4	4	15
—When interviewer arrived	3	4		8	2	17
Total estabs. interviewed	42	47	35	211	47	382
Total interviews (pairs)	65	55	43	237	52	452
Total pairs after D.P.	65	55	43	237	52	452
Total letters sent	49	89		223	53	414

For description of category inclusions, please see text.

TABLE C7—RESPONSE RATE ANALYSIS—ACTUAL FIGURES

UK (Wave 1)	RETAIL	BANKS	INSUR-ANCE	MECH. ENG.	ELEC-TRONICS	TOTAL
Initial addresses	151	260	319	750	368	1848
Not used (over quota)	—	—	—	—	—	—
Non-contacts (no business, etc.)	4	2	5	39	28	78
Non-contacts (respondent not available)	—	—	—	15	11	26
Non-eligible						
—Total	18	75	75	182	170	520
—Sector	—	—	—	—	64	64
—Size	4	36	40	51	31	162
—No new technology	8	2	8	48	24	90
—No Emp. Rep.	2	30	17	51	46	146
—No involvement	4	7	10	32	5	58
Pre-screening refusals	25	82	97	129	39	372
Post screening refusals	36	70	93	148	43	390
—At/during screening	23	45	68	128	26	290
—After letter received	10	19	18	17	8	72
—When interviewer arrived	3	6	7	3	9	28
Total estabs. interviewed	68	31	49	437	77	462
Total interviews (pairs)	72	31	51	257	82	493
Total pairs after D.P.	72	31	51	257	82	493
Total letters sent	71	56	74	257	94	534

For description of category inclusions, please see text.

TABLE C8—RESPONSE RATE ANALYSIS—ACTUAL FIGURES

BELGIUM (Wave 1)	RETAIL	BANKS	INSUR-ANCE	MECH. ENG.	ELEC-TRONICS	TOTAL
Initial addresses	285	112	116	362	227	1102
Not used (over quota)	132	16	—	—	—	148
Non-contacts (no business, etc.)	50	18	25	31	28	152
Non-contacts (respondent not available)	—	—	—	—	—	—
Non-eligible						
—Total	33	19	19	108	33	212
—Sector	2	—	—	25	4	31
—Size	18	11	17	61	15	122
—No new technology	2	—	—	11	4	17
—No Emp. Rep.	3	5	2	5	4	19
—No involvement	8	3	—	6	6	23
Pre-screening refusals	24	12	28	94	106	264
Post screening refusals	10	7	4	33	11	65
—At/during screening	2	—	—	3	—	5
—After letter received	—	5	3	13	3	24
—When interviewer arrived	8	2	1	17	8	36
Total estabs. interviewed	36	40	40	96	49	261
Total interviews (pairs)	46	61	51	124	70	352
Total pairs after D.P.	46	61	51	120	70	348
Total letters sent	46	47	44	129	60	326

For description of category inclusions, please see text.

TABLE C9—RESPONSE RATE ANALYSIS—ACTUAL FIGURES

GREECE (Wave 1)	RETAIL	BANKS	INSUR-ANCE	MECH. ENG.	ELEC-TRONICS	TOTAL
Initial addresses	100	350	83	67	53	653
Not used (over quota)	29	—	—	—	—	29
Non-contacts (no business, etc.)	6	3	13	4	1	27
Non-contacts (respondent not available)	—	—	—	—	—	—
Non-eligible						
—Total	11	189	45	14	16	275
—Sector	—	—	—	3	2	5
—Size	11	180	45	3	4	243
—No new technology	—	4	—	3	2	9
—No Emp. Rep.	—	5	—	4	7	16
—No involvement	—	—	—	1	1	2
Pre-screening refusals	4	3	—	6	—	13
Post screening refusals	14	135	5	35	7	196
—At/during screening	—	—	—	—	—	—
—After letter received	—	12	—	15	—	29
—When interviewer arrived	14	123	5	20	5	167
Total estabs. interviewed	36	20	20	8	29	113
Total interviews (pairs)	37	26	20	10	35	127
Total pairs after D.P.	37	26	20	10	34	127
Total letters sent	79	155	25	43	36	338

For description of category inclusions, please see text.

TABLE C10—RESPONSE RATE ANALYSIS—ACTUAL FIGURES

IRELAND (Wave 1)	RETAIL	BANKS	INSUR-ANCE	MECH. ENG.	ELEC-TRONICS	TOTAL
Initial addresses	94	61	125	47	113	440
Not used (over quota)	—	—	—	—	—	—
Non-contacts (no business, etc.)	2	1	8	3	2	16
Non-contacts (respondent not available)	—	—	—	—	—	—
Non-eligible						
—Total	63	44	105	25	86	323
—Sector	—	4	17	4	36	61
—Size	29	12	57	9	2	109
—No new technology	8	7	5	4	2	26
—No Emp. Rep.	9	11	23	8	16	67
—No involvement	17	10	3	—	30	60
Pre-screening refusals	6	3	3	2	4	18
Post screening refusals	10	8	6	7	14	45
—At/during screening	—	—	—	—	—	—
—After letter received	—	—	—	—	—	—
—When interviewer arrived	10	8	6	7	14	45
Total estabs. interviewed	13	5	3	10	7	38
Total interviews (pairs)	13	5	3	10	7	38
Total pairs after D.P.	13	5	3	10	7	38
Total letters sent	—	—	—	—	—	—

For description of category inclusions, please see text.

TABLE C11—RESPONSE RATE ANALYSIS—ACTUAL FIGURES

LUXEMBOURG (Wave 1)	RETAIL	BANKS	INSUR-ANCE	MECH. ENG.	ELEC-TRONICS	TOTAL
Initial addresses	10	84	14	44	20	172
Not used (over quota)	—	—	—	—	—	—
Non-contacts (no business, etc.)	—	—	—	1	—	1
Non-contacts (respondent not available)	—	—	—	—	—	—
Non-eligible						
—Total	3	21	10	20	14	68
—Sector	—	—	—	3	4	7
—Size	3	12	10	13	9	47
—No new technology	—	—	—	4	1	5
—No Emp. Rep.	—	9	—	—	—	9
—No involvement	—	—	—	—	—	—
Pre-screening refusals	1	27	1	7	2	38
Post screening refusals	1	11	—	3	1	16
—At/during screening	—	—	—	—	—	—
—After letter received	—	4	—	2	—	6
—When interviewer arrived	1	7	—	1	1	10
Total estabs. interviewed	5	25	3	13	3	49
Total interviews (pairs)	5	32	5	15	4	61
Total pairs after D.P.	5	32	5	15	4	61
Total letters sent	—	10	2	10	1	23

For description of category inclusions, please see text.

TABLE C12—RESPONSE RATE ANALYSIS—ACTUAL FIGURES

NETHERLANDS (Wave 2)	RETAIL	BANKS	INSUR-ANCE	MECH. ENG.	ELEC-TRONICS	TOTAL
Initial addresses	413	38	152	516	286	1705
Not used (over quota)	—	—	—	—	—	
Non-contacts (no business, etc.)	70	10	7	60	50	197
Non-contacts (respondent not available)	49	36	8	38	16	147
Non-eligible						
—Total	99	63	26	176	47	411
—Sector	21	12	3	49	18	103
—Size	39	42	15	88	2	186
—No new technology	12	—	2	10	9	33
—No Emp. Rep.	11	7	5	20	12	55
—No involvement	16	2	1	9	6	34
Pre-screening refusals	70	130	40	40	39	319
Post screening refusals	97	52	45	133	75	402
—At/during screening	57	34	31	94	44	260
—After letter received	23	6	10	29	30	98
—When interviewer arrived	17	12	4	10	1	44
Total estabs. interviewed	28	47	26	69	59	229
Total interviews (pairs)	28	55	29	79	68	259
Total pairs after D.P.	28	55	29	79	68	259
Total letters sent	68	65	40	108	90	371

For description of category inclusions, please see text.

TABLE C13—RESPONSE RATE ANALYSIS—ACTUAL FIGURES

PORTUGAL (Wave 2)	RETAIL	BANKS	INSUR-ANCE	MECH. ENG.	ELEC-TRONICS	TOTAL
Initial addresses	982	158	251	319	178	1888
Not used (over quota)	344	—	—	—	—	344
Non-contacts (no business, etc.)	—	—	—	8	3	11
Non-contacts (respondent not available)	—	—	—	—	—	—
Non-eligible						
—Total	578	135	229	261	124	1327
—Sector	49	3	9	44	17	122
—Size	178	118	208	17	—	521
—No new technology	96	—	1	44	10	151
—No Emp. Rep.	207	1	1	103	62	374
—No involvement	48	13	10	53	35	159
Pre-screening refusals	—	—	—	—	—	—
Post screening refusals	32	9	8	17	22	88
—At/during screening	6	1	1	7	10	25
—After letter received	5	6	2	3	3	19
—When interviewer arrived	21	2	5	7	9	44
Total estabs. interviewed	28	14	14	33	29	118
Total interviews (pairs)	50	25	23	61	47	206
Total pairs after D.P.	50	24	23	57	46	200
Total letters sent	54	22	21	43	41	181

For description of category inclusions, please see text.

TECHNICAL APPENDIX

TABLE C14—RESPONSE RATE ANALYSIS—ACTUAL FIGURES

SPAIN (Wave 2)	RETAIL	BANKS	INSUR-ANCE	MECH. ENG.	ELEC-TRONICS	TOTAL
Initial addresses	160	201	183	608	352	1504
Not used (over quota)	—	—	—	—	—	—
Non-contacts (no business, etc.)	10	5	4	20	19	58
Non-contacts (respondent not available)	17	7	4	1	2	31
Non-eligible						
—Total	41	62	64	415	107	689
—Sector	7	—	5	107	32	151
—Size	32	61	58	300	63	514
—No new technology	2	1	1	8	12	24
—No Emp. Rep.	—	—	—	—	—	—
—No involvement	—	—	—	—	—	—
Pre-screening refusals	20	30	20	10	11	91
Post screening refusals	37	46	54	113	138	388
—At/during screening	22	21	30	89	92	254
—After letter received	5	7	10	12	39	73
—When interviewer arrived	10	18	14	12	7	61
Total estabs. interviewed	35	51	37	49	75	247
Total interviews (pairs)	40	63	44	60	96	303
Total pairs after D.P.	40	63	44	60	96	303
Total letters sent	50	76	61	73	121	381

For description of category inclusions, please see text.

3.2 Response rate analysis

In order to arrive at an analysis of response rates, it is first of all necessary to calculate the universe of eligible establishments for each country. As mentioned above, the major problem in doing this is the pre-screening refusals where eligibility was never determined. Assuming these establishments would split in exactly the same proportions as the screened sample, then it is possible to arrive at an estimate of eligible establishments. Using Belgium as an example, the following is an example calculation of the method used for the response rate analysis by country on Table D:

STAGE 1: Eligibility calculation (Belgium)

Starting address   1102

Minus: non-used sample   148 ⎫
        + non-contactable   152 ⎬ ⇒ 564
        + screening refusals   264 ⎭

Leaves screened sample of   538
Non-eligibles   =   212   (39.4% of screened sample)
Pro-rata screening refusals (264)= 104   non-eligible
                                   260   eligible

STAGE 2: Response rate calculation

Starting addresses   1102

Minus: Not used samples   148 ⎫
        + non-contactables   152 ⎬ ⇒ 616 (56% of total
        + non-eligibles (screened)   212 ⎬   addresses)
        + non-eligibles (non-screened)104 ⎭

Leaves: eligible sample of   486

Fieldwork achievement equals:

Pre-screening refusals (estimated):   160 (33%)
Post-screening refusals: 65 (13%)
No. of establishments interviewed   261 (54%)

TECHNICAL APPENDIX

TABLE D—RESPONSE RATE ANALYSIS—BY ESTABLISHMENT
(see text for calculation method)

WAVE 1	DENMARK	FRANCE	GERMANY	ITALY	U.K.	TOTAL
Initial addresses	440	2369	4283	3381	1848	12321
Not used/non-eligible	120	1146	3059	2721	765	7814
% of total	27%	48%	71%	80%	41%	63%
Eligible establishments	320	1228	1224	660	1083	4510
% Response						
—Pre-screening refusals	—	19%	14%	7%	21%	15%
—Post screening refusals	35%	52%	43%	35%	36%	42%
—Interviewed establishments	65%	29%	43%	58%	43%	43%

WAVE 2	BELG.	GREECE	IREL.	LUX.	NETHS.	PORT.	SPAIN	TOTAL
Initial addresses	1102	1653	440	172	1705	1888	1504	7464
Not used/non-eligible	616	337	353	88	881	1682	822	4779
% of total	56%	52%	80%	51%	52%	89%	55%	64%
Eligible establishments	486	316	87	84	824	206	682	2685
% Response								
—Pre-screening refusals	33%	4%	4%	23%	23%	—	7%	16%
—Post screening refusals	13%	62%	52%	19%	49%	43%	57%	45%
—Interviewed establishments	54%	36%	44%	58%	28%	57%	36%	39%

3.3 Commentary on response rates

In all countries in both waves, the survey procedure was a complex one, because of the number of different levels of respondent requiring to be contacted. To support the approach, co-operation was obtained from the Foundation Administration Board members in each country to respondents. This helped to reassure them that the survey was a genuine and worthwhile exercise.

However, even with the full co-operation of the Board, difficulties were experienced in completing the survey in some countries, notably in wave 2. Above all, the large number of establishments which proved to be ineligible severely limited the establishment universe. 5 countries with the exception of the Netherlands failed to complete the target set, (Denmark in Wave 1, Greece, Ireland, Luxembourg, Netherlands in Wave 2). In all of these countries, eligibility barriers appeared on other criteria, notably lack of new technology or lack of formal employee representation, since the smaller the establishment, the less likely it was that these criteria would be met.

Apart from the contribution to non-response resulting from the industrial climate, the structure of the survey design also contributed in some respects. The following were the main problems met:

— The starting point of the survey was the personnel manager, the only practical way of getting into companies. However, in some instances, this approach resulted in wrong names being given of potential respondents, causing difficulties in tracing them.

— Additionally, this approach sometimes resulted in suspicion or resentment on the part of nominated respondents, who then had to be persuaded or reassured about the survey value, but who ultimately refused to participate.

— Some refusals at branch level occurred because staff felt that the issues raised in the survey were more appropriately dealt with at Head Office level.

— At line manager level, some objections were raised to the relevant employee representative being interviewed on the grounds of the sensitivity of the subject matter. In some smaller companies, the line manager objected on the grounds of wasting the employee representative's time.

4. Fieldwork Dates

WAVE 1: Between 11th February and 22nd May 1987

WAVE 2: Between 5th April and 7th October 1988

5. Weighting Procedures

The weighting procedure used has been designed to take into account a) the relative importance of each sector in each country and b) the relative size of each of the 12 countries to each other.

It would have been preferable to weight the data to the establishment universe for Europe, but unfortunately there is no accurate information source available for all countries. Data are therefore weighted to the number of employees in all countries, using the Arbeitskosten Stude 1984, published by Eurostat.

The procedure adopted was as follows:

(a) Split the universe of employees in each country for the sectors covered in the survey into size bank groupings by establishing the number of employees in each sector within each size band.
(N.B. Size bands vary by sector and by country, reflecting the differences in industry structures in each country). (Table E).

(b) Determine the actual proportion of total employees in each of these sector size band groupings in each country. This becomes the target weight for each cell. (Table E).

(c) Establish the number of employees in all the companies interviewed, and determine the proportions in each of the sector size band groupings. (Table F).

(d) Each of the resulting cells is then weighted to the cell targets calculated in step b.

(e) The result is a 5 sector x12 country matrix, where each cell has a sector weight and a country weight relative to the size of each country. (Table G).

The following tables show the weighting matrices used in the survey weighting.

TABLE E

EMPLOYEE UNIVERSE BY SECTOR

EMPLOYEES UNIVERSE—RETAIL

ABSOLUTE FIGURES
PERCENTAGES

RETAIL	UNITED KINGDOM	FRANCE	GERMANY	ITALY	DENMARK	IRELAND	BELGIUM	NETHER-LANDS	SPAIN	PORTUGAL	GREECE	LUXEMBURG
10-49											3,426 *7.10*	
10-99						2,416 *4.48*	15,893 *6.44*					
10-199									52,906 *13.21*	9,487 *7.12*		
50-99		30,794 *2.21*		4,061 *0.52*	2,151 *1.70*							
50-199	41,366 *2.00*		38,067 *1.51*					10,580 *3.21*			1,756 *3.63*	860 *6.13*
50+				3,227 *0.41*								
100-199												
100-499		45,360 *3.25*										
100+					4,327 *3.41*	12,685 *23.54*	24,268 *9.84*					
200-499	28,229 *1.37*		36,806 *1.46*									
200+				29,670 *3.78*				55,112 *16.74*	31,248 *7.80*	4,463 *3.22*		
500+	349,363 *16.92*	217,195 *15.55*	236,436 *9.36*									

EMPLOYEES UNIVERSE—BANKING

ABSOLUTE FIGURES
PERCENTAGES

BANKING	UNITED KINGDOM	FRANCE	GERMANY	ITALY	DENMARK	IRELAND	BELGIUM	NETHER-LANDS	SPAIN	PORTUGAL	GREECE	LUXEMBURG
10-49												1,480 *10.54*
10-99												
10-199										56,059 *40.50*		
50-99				9,471 *1.21*	2,149 *1.70*	1,215 *2.26*	1,773 *0.72*	6,722 *2.04*				
50-199	4,705 *0.23*	27,592 *1.98*	90,115 *3.57*						41,213 *10.30*	6,760 *13.99*		
50+												7,519 *53.56*
100-199				13,903 *1.77*	1,962 *1.55*			7,262 *2.21*				
100-499							8,067 *3.30*					
100+						18,994 *35.25*						
200-499	10,238 *0.50*	53,579 *3.84*	86,030 *3.41*						29,836 *7.48*			
200+				289,738 *36.93*	42,385 *33.44*			70,727 *21.48*			8,699 *18.00*	
500+												
1000+												

TECHNICAL APPENDIX

EMPLOYEES UNIVERSE—INSURANCES

ABSOLUTE FIGURES
PERCENTAGES

INSURANCES	UNITED KINGDOM	FRANCE	GERMANY	ITALY	DENMARK	IRELAND	BELGIUM	NETHER-LANDS	SPAIN	PORTUGAL	GREECE	LUXEMBURG
10-99											3,168 *6.56*	
50-99					783 *0.62*				5,582 *1.39*			
50-199	3,877 *0.19*	11,617 *0.83*	9,298 *0.37*	6,591 *0.84*			4,393 *1.78*	5,648 *1.72*				
50-499										3,578 *2.58*		438 *3.12*
50+												
100-199					1,251 *0.99*				6,410 *1.60*			
100-499											1,068 *2.21*	
100+						7,856 *14.58*						
200-499	12,008 *0.58*	13,723 *0.98*	21,248 *0.84*				8,167 *3.30*					
200+				35,557 *4.53*	12,728 *10.04*			29,548 *8.97*	6,221 *1.55*		1,458 *3.02*	
500-999												
500+	143,748 *6.96*	86,097 *6.16*	169,111 *6.69*				11,454 *4.64*			9,666 *6.99*		

TECHNICAL APPENDIX

EMPLOYEES UNIVERSE—MECHANICAL ENGINEERING

ABSOLUTE FIGURES
PERCENTAGES

MECHANICAL ENGINEERING	UNITED KINGDOM	FRANCE	GERMANY	ITALY	DENMARK	IRELAND	BELGIUM	NETHER-LANDS	SPAIN	PORTUGAL	GREECE	LUXEMBURG
10-49												154 *1.10*
10-99							12,537 *5.95*			6,837 *4.96*		
10+											6,355 *13.60*	
50-99					2,866 *2.26*			12,007 *3.65*				
50-199					19,620 *10.15*				41,237 *10.29*			
50+												3,003 *21.39*
100-199	83,560 *4.04*	63,387 *4.54*	113,671 *4.50*		6,280 *4.96*			12,702 *3.86*		7,448 *5.39*		
100-499												
100+						7,856 *14.58*	27,157 *11.01*					
200-499	134,245 *6.55*	57,877 *4.14*	184,514 *7.30*	49,384 *6.29*					23,378 *5.84*			
200+					32,489 *25.64*			12,755 *3.87*		7,990 *5.78*		
500-999	97,937 *4.74*	52,562	389,757	65,741								
500+									15,247 *3.80*			
1000+	157,893 *7.65*											

271

EMPLOYEES UNIVERSE—ELECTRONIC ENGINEERING

ABSOLUTE FIGURES
PERCENTAGES

ELECTRONIC ENGINEERING	UNITED KINGDOM	FRANCE	GERMANY	ITALY	DENMARK	IRELAND	BELGIUM	NETHER-LANDS	SPAIN	PORTUGAL	GREECE	LUXEMBURG
10-99							8,024 3.25			3,393 2.45	5,589 11.59	
10+												584 4.16
50-99		35,246 2.52		15,546 1.98	1,816 1.43			3,492 1.06	9,823 2.45			
50-199	108,548 5.26		106,362 4.21									
50+						6,330 11.75						
100-199		59,220 4.24		20,976 2.76	4,202 3.31		3,076 1.25	4,343 1.32	14,680 3.60		2,628 5.44	
100-499										9,068 6.56		
100+												
200-499	111,717 5.14		114,069 4.52				9,352 3.79		29,105 7.31			
200+		336,471 24.09		161,144 20.54	11,342 8.95			98,362 29.87			7,388 15.30	
500-999												
500+	335,826 16.27		661,194 26.17				47,471 19.24		76,144 19.00	19,919 14.45		
1000+												

TABLE F

EMPLOYEE SAMPLE BY SECTOR

EMPLOYEES SAMPLE—RETAIL

ABSOLUTE FIGURES
PERCENTAGES

RETAIL	UNITED KINGDOM	FRANCE	GERMANY	ITALY	DENMARK	IRELAND	BELGIUM	NETHER-LANDS	SPAIN	PORTUGAL	GREECE	LUXEMBURG
10-49											626 / 2.68	
10-99						402 / 4.55	748 / 0.52					
10-199									2,119 / 2.22	4,914 / 9.58		
50-99		701 / 0.35		1,917 / 1.88	1,517							
50-199	4,638 / 2.09		2,836 / 0.92					1,911 / 3.47				
50+											539 / 2.30	1,173 / 6.44
100-199				3,312 / 1.70								
100-499		21,250 / 10.61										
100+			4,236 / 1.37		4,283 / 5.31	2,906 / 32.90	10,433 / 7.24					
200-499	5,756 / 2.53			6,770 / 3.48								
200+								3,761 / 6.83	20,045 / 21.03	2,618 / 5.10		
500+	7,230 / 3.18	25,564 / 12.77	14,519 / 4.70									

274

TECHNICAL APPENDIX

EMPLOYEES UNIVERSE—BANKING

ABSOLUTE FIGURES
PERCENTAGES

BANKING	UNITED KINGDOM	FRANCE	GERMANY	ITALY	DENMARK	IRELAND	BELGIUM	NETHER-LANDS	SPAIN	PORTUGAL	GREECE	LUXEMBURG
10-49												144 *0.80*
10-99												
10+										14,880 *28.99*		
50-99				450 *0.23*	1,309 *1.62*	160 *1.81*	653 *0.45*	1,157 *2.10*				
50-199	1,478 *0.65*	2,892 *1.45*	5,989 *1.94*						4,062 *4.26*		1,549 *6.64*	11,439 *62.80*
50+												
100-199				2,266 *1.17*	1,294 *1.61*			2,929 *5.32*				
100-499							4,705 *3.26*					
100+						1,785 *20.19*						
200-499	2,400 *1.05*	10,619 *5.30*	7,133 *2.31*									
200+				23,342 *12.00*	5,050 *6.27*			3,985 *7.23*	4,632 *4.86*		3,374 *14.50*	
500-999												
500+	6,400 *2.81*	37,031 *18.49*	17,842 *5.77*				49,402 *34.26*		13,577 *14.25*			
1000+												

275

EMPLOYEES UNIVERSE—INSURANCES

ABSOLUTE FIGURES
PERCENTAGES

INSURANCES	UNITED KINGDOM	FRANCE	GERMANY	ITALY	DENMARK	IRELAND	BELGIUM	NETHER-LANDS	SPAIN	PORTUGAL	GREECE	LUXEMBURG
10-99												321 *1.38*
50-99					210 *0.26*				203 *0.21*			
50-199	2,544 *1.12*	1,608 *0.80*	2,982 *0.96*	1,885 *0.97*			2,856 *1.98*	1,702 *3.09*				
50-499												
50+										3,165 *6.17*		960 *5.27*
100-199					910 *1.13*				3,557 *3.73*		1,470 *6.28*	
100-499												
100+						361 *4.09*						
200-499	4,386 *1.93*	6,466 *3.23*	5,040 *1.63*				6,563 *4.55*					
200+				40,490 *20.82*	11,480 *14.24*			6,578 *11.98*	4,979 *5.22*		1,289 *5.53*	
500-999												
500+	9,457 *4.15*	20,873 *10.42*	46,122 *14.93*				11,376 *7.89*	10.26		5,266		
1000+												

276

TECHNICAL APPENDIX

EMPLOYEES UNIVERSE—MECHANICAL ENGINEERING

ABSOLUTE FIGURES
PERCENTAGES

MECHANICAL ENGINEERING	UNITED KINGDOM	FRANCE	GERMANY	ITALY	DENMARK	IRELAND	BELGIUM	NETHER-LANDS	SPAIN	PORTUGAL	GREECE	LUXEMBURG
10-49												79 / *0.43*
10-99							1,695 / *1.18*			1,726 / *3.36*		
10+											9,015 / *38.67*	
50-99								2,500 / *4.54*				
50-199				15,273 / *7.86*								
50+						818 / *9.26*			4,095 / *4.30*			3,652 / *20.04*
100-199	13,510 / *5.94*	9,146 / *4.56*	14,375 / *4.86*		7,213 / *8.95*		5,959 / *4.11*	4,698 / *8.54*		2,780 / *5.42*		
100-499												
100+						7,856 / *14.58*						
200-499	27,732 / *12.19*	13,411 / *6.70*	27,046 / *8.75*	24,432 / *12.57*					7,105 / *7.46*			
200+					25,803 / *32.02*		25,298 / *17.55*	8,053 / *14.63*		7,280 / *14.20*		
500-999	30,919 / *13.59*											
500+		8,100 / *4.04*	104,754 / *33.90*	51,230 / *26.35*					1,929			
1000+	43,179 / *18.97*											

EMPLOYEES UNIVERSE—ELECTRONIC ENGINEERING

ABSOLUTE FIGURES
PERCENTAGES

ELECTRONIC ENGINEERING	UNITED KINGDOM	FRANCE	GERMANY	ITALY	DENMARK	IRELAND	BELGIUM	NETHER-LANDS	SPAIN	PORTUGAL	GREECE	LUXEMBURG
10-99							1,505 / 1.04			1,156 / 2.25	949 / 4.07	
10+												769 / 4.22
50-99		2,555 / 1.28		763 / 0.39	1,056 / 1.31			1,831 / 3.32	2,804 / 2.94			
50-199	1,281 / 0.56		1,778 / 0.58									
50+						7,407 / 27.20						
100-199		4,428 / 2.21		2,497 / 1.28	1,118 / 1.39		2,403 / 1.67	1,979 / 3.60	2,687 / 2.82		892 / 3.83	
100-499										3,438 / 6.69		
100+												
200-499	7,767 / 3.41		5,796 / 1.88				4,633 / 3.21		6,749 / 7.08			
200+		35,632 / 17.79		19,804 / 10.19	15,534 / 19.28			13,945 / 25.34			3,291 / 14.12	
500-999												
500+	58,892 / 25.88		48,546 / 15.71				15,974 / 11.09		16,757 / 17.60	4,093 / 7.98		
1000+												

TABLE G

EMPLOYEE SAMPLE BY SECTOR

WEIGHTING FIGURES—RETAIL

COUNTRY WEIGHTS
EUROPE WEIGHTS

RETAIL	UNITED KINGDOM	FRANCE	GERMANY	ITALY	DENMARK	IRELAND	BELGIUM	NETHER-LANDS	SPAIN	PORTUGAL	GREECE	LUXEMBURG
10-49											2.649 *0.940*	
10-99						0.985 *1.032*	12.384 *3.678*					
10-199									5,950 *4.322*	0.743 *0.347*		
50-99		6.314 *7.627*	0.276	0.525 *0.247*	0.904							
50-199	0.980 *1.541*	2.322	1.641					0.925 *0.958*				
50+				0.241 *0.168*							1.578 *0.560*	0.952 *0.126*
100-199												
100-499		0.306 *0.370*										
100+					0.642 *0.175*	0.716 *0.750*	1.359 *0.404*					
200-499	0.542 *0.852*		1.066 *1.508*									
200+				1.086 *0.759*				2.451 *2.539*	0.371 *0.270*	0.631 *0.295*		
500+	5.321 *8.365*	1.218 *1.471*	1.991 *2.817*									

280

WEIGHTING FIGURES—BANKING

COUNTRY WEIGHTS
EUROPE WEIGHTS

BANKING	UNITED KINGDOM	FRANCE	GERMANY	ITALY	DENMARK	IRELAND	BELGIUM	NETHER-LANDS	SPAIN	PORTUGAL	GREECE	LUXEMBURG
10-49												13.175 *1.739*
10-99												
10+										1.397 *0.652*		
50-99				5.261 *3.677*	1.049 *0.286*	1.249 *1.309*	1.600 *0.475*	0.971 *1.006*				
50-199	0.354 *0.556*	1.336 *1.614*	1.840 *2.604*						2.418 *1.760*		2.107 *0.748*	
50+												0.853 *0.113*
100-199				1.513 *1.058*	0.963 *0.263*			0.415 *0.430*				
100-499							1.012 *0.301*					
100+						1.746 *1.830*						
200-499	0.476 *0.748*	0.725 *0.876*	1.476 *2.089*						1.539 *1.120*			
200+				3.078 *2.152*	5.333 *1.456*			2.971 *3.078*				
500-999												
500+	7.591 *11.933*	1.184 *1.430*	1.847 *2.614*				0.692 *0.206*		0.307 *0.223*		1.241 *0.441*	
1000+												

281

WEIGHTING FIGURES—INSURANCES

COUNTRY WEIGHTS
EUROPE WEIGHTS

INSURANCES	UNITED KINGDOM	FRANCE	GERMANY	ITALY	DENMARK	IRELAND	BELGIUM	NETHER-LANDS	SPAIN	PORTUGAL	GREECE	LUXEMBURG
10-99											4.754 *1.688*	
50-99					2.385 *0.651*				6.619 *4.819*			
50-199	0.170 *0.267*	1.038 *1.254*	0.385 *0.545*	0.866 *0.605*			0.899 *0.267*					
50-499										0.418 *0.195*		
50+												0.592 *0.078*
100-199					0.876 *0.239*			0.557 *0.577*	0.429 *0.312*		0.352 *0.125*	
100-499												
100+						3.565 *3.736*						
200-499	0.301 *0.473*	0.303 *0.366*	0.515 *0.729*	0.218 *0.152*			0.725 *0.215*					
200+					0.705 *0.92*			0.749 *0.776*	0.297 *0.216*		0.546 *0.194*	
500-999												
500+	1.677 *2.636*	0.591 *0.714*	0.448 *0.634*				0.588 *0.175*			0.681 *0.318*		
1000+												

TECHNICAL APPENDIX

WEIGHTING FIGURES—MECHANICAL ENGINEERING

COUNTRY WEIGHTS
EUROPE WEIGHTS

MECHANICAL ENGINEERING	UNITED KINGDOM	FRANCE	GERMANY	ITALY	DENMARK	IRELAND	BELGIUM	NETHER-LANDS	SPAIN	PORTUGAL	GREECE	LUXEMBURG
10-49												2.558 *0.338*
10-99							4.279 *1.271*			1.476 *0.689*		
10+											0.340 *0.121*	
50-99					0.478 *0.130*			0.804 *0.833*				
50-199				1.291 *0.002*					2.393 *1.742*			
50+						0.879 *0.921*						1.067 *0.137*
100-199	0.680 *1.069*	0.996 *1.203*	0.968 *1.370*		0.554 *0.151*		0.633 *0.188*	0.452 *0.468*		0.994 *0.464*		
100-499												
100+												
200-499	0.537 *0.844*	0.618 *0.747*	0.834 *1.180*	0.500 *0.350*					0.783 *0.570*	0.407 *0.190*		
200+					0.801 *0.219*		0.627 *0.186*	0.265 *0.275*				
500-999												
500+		0.933 *1.127*	0.455 *0.644*	0.318 *0.222*					1.881 *1.369*			
1000+	0.403 *0.216*											

283

WEIGHTING FIGURES—ELECTRONIC ENGINEERING

ELECTRONIC ENGINEERING	UNITED KINGDOM	FRANCE	GERMANY	ITALY	DENMARK	IRELAND	BELGIUM	NETHER-LANDS	SPAIN	PORTUGAL	GREECE	LUXEMBURG
10-99							3.125 / 0.928			1.089 / 0.508	2.847 / 1.011	
10+												0.986 / 0.130
50-99		1.969 / 2.379		5.077 / 3.549	1.092 / 0.298			0.319 / 0.330	0.833 / 0.606			
50-199		9.393 / 14,766		7,258 / 10,270								
50+						0.437 / 0.453						
100-199		1.919 / 2.318		2.086 / 1.458	2.381 / 0.650		0.748 / 0.222	0.367 / 0.380	1.277 / 0.930		1.420 / 0.504	
100-499										0.981 / 0.458		
100+												
200-499	1.587 / 2.495		2.404 / 3.402				1.181 / 0.351		1.032 / 0.751			
200+		1.354 / 1.636		2.016 / 1.409	0.464 / 0.127			1.179 / 1.221			1.083 / 0.384	
500-999												
500+	0.629 / 0.989		1.666 / 2.357				1.735 / 0.515		1.079 / 0.786	1.811 / 0.846		
1000+												

COUNTRY WEIGHTS / *EUROPE WEIGHTS*

6. Agencies involved

The Harris Research Centre, London and GfK Marketforschung, Nuremberg were the agencies responsible for co-ordinating the survey, commissioning sub-contracting agencies to carry out fieldwork in each of the 12 European countries. The following were the agencies used:

WAVE 1
Denmark	Observa, Vedbaek
Italy	ASM, Rome
France	Louis Harris, Paris
Germany	GfK, Nuremberg
UK	Harris Research, London

WAVE 2
Belgium	Sobemap, Brussels
Greece	Nielsen, Athens
Ireland	IMS, Dublin
Luxembourg	Ilres, Luxembourg
Netherlands	Nipo, Amsterdam
Portugal	Nielsen, Lisbon
Spain	Dym, Barcelona

7. Questionnaire translations

An English master version of the questionnaire was prepared, and sent to each agency for translation into national languages.

Each translated questionnaire was then sent back to Harris Research for checking. This was done by sending the translations to foreign nationals resident in the UK who then back-translated them back into English. These back-translations were then checked word for word against the original version to identify any inconsistencies and amended as necessary.

Questionnaires were modified slightly for each sector, to take account of differences between them. In total, the following were used:

A screening questionnaire — 5 versions, one for each sector
A company questionnaire — 5 versions, one for each sector

A master version of the questionnaires used (in English) is appended.

8. Levels of technology by sector

The following technologies were specified as eligible for inclusion in the survey.

Retailing
  (i) Electronic Cash Registers
      Head Office Mainframe Computer
      In-store Mini-Computer: Stand Alone

  (ii) Stand Alone Data Capture terminals
      Visual Display Unit (VDU) terminals linked to mainframe computer

  (iii) Stand Alone Point of Sales (POS) terminals
      On-line POS terminals or networked to computer
      Electronic Fund Transfer
      In-store Mini-Computer: On-line
      Computerised Store Control

Banking and Insurance
  (i) Head Office Mainframe computer
      Automatic teller machines
      Wordprocessing facilities
      Visual Display Unit (VDU) terminals for data access
      Optical Character Recognition

  (ii) VDU terminals for data capture/interactive systems, but with processing of transaction administered elsewhere

  (iii) Cashier terminals
      Minicomputers in branch offices
      VDU terminals to administer complete transactions in branch offices
      End User computing facilities
      Agent or broker terminals with direct computer link to other companies

  (iv) Integration of word-processing and dataprocessing
      Electronic mail-box
      Electronic filing

Mechanical Engineering
 (i) Computer Numerical Control (CNC) machine Tools
    New 'Machining' Techniques
    Automated Materials Handling
    Automated Assembly
    Computer Aided Design
    Word Processing

 (ii) Computerised Inspection and Testing
    Computerised Stock or Production Control-based systems
    Computer-based office and administration systems not covered elsewhere.

 (iii) Flexible Manufacturing Cells
    Direct Numerical Control
    Computer Aided Design and Computer Aided Manufacture (CAD/CAM): use of the output of CAD for preparation of CNC tapes or manufacturing instructions
    Manufacturing Requirement Planning Systems

 (iv) Computer Integrated Manufacture

Electronics
 (i) Computer Numerical Control (CNC) machine tools
    Automated materials handling
    Automatic Component Insertion
    Automated assembly
    Computer Aided Design
    Word Processing

 (ii) Computerised Inspection and Testing
    Computerised Stock or Production Control—Basic systems
    Computer-based office and administration systems not covered elsewhere

 (iii) Flexible Manufacturing Cells
    Computer Aided Design and Computer Aided Manufacture (CAD/CAM): use of the output of CAD for preparation of CNC tapes or manufacturing instruction
    Manufacturing Requirement Planning Systems

 (iv) Computer Integrated Manufacture

European Foundation for the Improvement of Living and Working
Conditions

**Workplace Involvement in Technological Innovation in the European
Community**
Volume I: Roads to Participation

Luxembourg: Office of Official Publications of the
European Communities

1993—300 p.—16 × 23.5 cm

ISBN 92-826-5669-1 (Vol. I)

ISBN 92-826-6026-5 (Vol. I and II)

Price (excluding VAT) in Luxembourg:

Vol. I: ECU 31,50
Vol. I and II: ECU 54

Venta y suscripciones • Salg og abonnement • Verkauf und Abonnement • Πωλήσεις και συνδρομές
Sales and subscriptions • Vente et abonnements • Vendita e abbonamenti
Verkoop en abonnementen • Venda e assinaturas

BELGIQUE / BELGIË

**Moniteur belge /
Belgisch Staatsblad**

Rue de Louvain 42 / Leuvenseweg 42
B-1000 Bruxelles / B-1000 Brussel
Tél. (02) 512 00 26
Fax (02) 511 01 84

Autres distributeurs /
Overige verkooppunten

**Librairie européenne /
Europese boekhandel**

Rue de la Loi 244 / Wetstraat 244
B-1040 Bruxelles / B-1040 Brussel
Tél. (02) 231 04 35
Fax (02) 735 08 60

Jean de Lannoy

Avenue du Roi 202 / Koningslaan 202
B-1060 Bruxelles / B-1060 Brussel
Tél. (02) 538 51 69
Télex 63220 UNBOOK B
Fax (02) 538 08 41

Document delivery:

Credoc

Rue de la Montagne 34 / Bergstraat 34
Bte 11 / Bus 11
B-1000 Bruxelles / B-1000 Brussel
Tél. (02) 511 69 41
Fax (02) 513 31 95

DANMARK

J. H. Schultz Information A/S

Herstedvang 10-12
DK-2620 Albertslund
Tlf. 43 63 23 00
Fax (Sales) 43 63 19 69
Fax (Management) 43 63 19 49

DEUTSCHLAND

Bundesanzeiger Verlag

Breite Straße 78-80
Postfach 10 80 06
D-W-5000 Köln 1
Tel. (02 21) 20 29-0
Telex ANZEIGER BONN 8 882 595
Fax 2 02 92 78

GREECE/ΕΛΛΑΔΑ

G.C. Eleftheroudakis SA

International Bookstore
Nikis Street 4
GR-10563 Athens
Tel. (01) 322 63 23
Telex 219410 ELEF
Fax 323 98 21

ESPAÑA

Boletín Oficial del Estado

Trafalgar, 29
E-28071 Madrid
Tel. (91) 538 22 95
Fax (91) 538 23 49

Mundi-Prensa Libros, SA

Castelló, 37
E-28001 Madrid
Tel. (91) 431 33 99 (Libros)
     431 32 22 (Suscripciones)
     435 36 37 (Dirección)
Télex 49370-MPLI-E
Fax (91) 575 39 98

Sucursal:

Librería Internacional AEDOS

Consejo de Ciento, 391
E-08009 Barcelona
Tel. (93) 488 34 92
Fax (93) 487 76 59

**Llibreria de la Generalitat
de Catalunya**

Rambla dels Estudis, 118 (Palau Moja)
E-08002 Barcelona
Tel. (93) 302 68 35
     302 64 62
Fax (93) 302 12 99

FRANCE

**Journal officiel
Service des publications
des Communautés européennes**

26, rue Desaix
F-75727 Paris Cedex 15
Tél. (1) 40 58 75 00
Fax (1) 40 58 77 00

IRELAND

Government Supplies Agency

4-5 Harcourt Road
Dublin 2
Tel. (1) 61 31 11
Fax (1) 78 06 45

ITALIA

Licosa SpA

Via Duca di Calabria, 1/1
Casella postale 552
I-50125 Firenze
Tel. (055) 64 54 15
Fax 64 12 57
Telex 570466 LICOSA I

GRAND-DUCHÉ DE LUXEMBOURG

Messageries du livre

5, rue Raiffeisen
L-2411 Luxembourg
Tél. 40 10 20
Fax 40 10 24 01

NEDERLAND

SDU Overheidsinformatie

Externe Fondsen
Postbus 20014
2500 EA's-Gravenhage
Tel. (070) 37 89 911
Fax (070) 34 75 778

PORTUGAL

Imprensa Nacional

Casa da Moeda, EP
Rua D. Francisco Manuel de Melo, 5
P-1092 Lisboa Codex
Tel. (01) 69 34 14

**Distribuidora de Livros
Bertrand, Ld.ª**

Grupo Bertrand, SA

Rua das Terras dos Vales, 4-A
Apartado 37
P-2700 Amadora Codex
Tel. (01) 49 59 050
Telex 15798 BERDIS
Fax 49 60 255

UNITED KINGDOM

HMSO Books (Agency section)

HMSO Publications Centre
51 Nine Elms Lane
London SW8 5DR
Tel. (071) 873 9090
Fax 873 8463
Telex 29 71 138

ÖSTERREICH

**Manz'sche Verlags-
und Universitätsbuchhandlung**

Kohlmarkt 16
A-1014 Wien
Tel. (0222) 531 61-0
Telex 112 500 BOX A
Fax (0222) 531 61-39

SUOMI/FINLAND

Akateeminen Kirjakauppa

Keskuskatu 1
PO Box 128
SF-00101 Helsinki
Tel. (0) 121 41
Fax (0) 121 44 41

NORGE

Narvesen Info Center

Bertrand Narvesens vei 2
PO Box 6125 Etterstad
N-0602 Oslo 6
Tel. (22) 57 33 00
Telex 79668 NIC N
Fax (22) 68 19 01

SVERIGE

BTJ

Tryck Traktorwägen 13
S-222 60 Lund
Tel. (046) 18 00 00
Fax (046) 18 01 25
         30 79 47

SCHWEIZ / SUISSE / SVIZZERA

OSEC

Stampfenbachstraße 85
CH-8035 Zürich
Tel. (01) 365 54 49
Fax (01) 365 54 11

ČESKÁ REPUBLIKA

NIS ČR

Havelkova 22
130 00 Praha 3
Tel. (2) 235 84 46
Fax (2) 235 97 88

MAGYARORSZÁG

Euro-Info-Service

Club Sziget
Margitsziget
1138 Budapest
Tel./Fax 1 111 60 61
         1 111 62 16

POLSKA

Business Foundation

ul. Krucza 38/42
00-512 Warszawa
Tel. (22) 21 99 93, 628-28 82
International Fax & Phone
(0-39) 12-00-77

ROMÂNIA

Euromedia

65, Strada Dionisie Lupu
70184 Bucuresti
Tel./Fax 0 12 96 46

BĂLGARIJA

Europress Klassica BK Ltd

66, bd Vitosha
1463 Sofia
Tel./Fax 2 52 74 75

RUSSIA

Europe Press

20, Sadovaja-Spasskaja Street
107078 Moscow
Tel. 095 208 28 60
     975 30 09
Fax 095 200 22 04

CYPRUS

**Cyprus Chamber of Commerce and
Industry**

Chamber Building
38 Grivas Dhigenis Ave
3 Deligiorgis Street
PO Box 1455
Nicosia
Tel. (2) 449500/462312
Fax (2) 458630

TÜRKIYE

**Pres Gazete Kitap Dergi
Pazarlama Dağitim Ticaret ve sanayi
AŞ**

Narlibahçe Sokak N. 15
Istanbul-Cağaloğlu
Tel. (1) 520 92 96 - 528 55 66
Fax 520 64 57
Telex 23822 DSVO-TR

ISRAEL

ROY International

PO Box 13056
41 Mishmar Hayarden Street
Tel. Aviv 61130
Tel. 3 496 108
Fax 3 544 60 39

UNITED STATES OF AMERICA/
CANADA

UNIPUB

4611-F Assembly Drive
Lanham, MD 20706-4391
Tel. Toll Free (800) 274 4888
Fax (301) 459 0056

CANADA

Subscriptions only
Uniquement abonnements

Renouf Publishing Co. Ltd

1294 Algoma Road
Ottawa, Ontario K1B 3W8
Tel. (613) 741 43 33
Fax (613) 741 54 39
Telex 0534783

AUSTRALIA

Hunter Publications

58A Gipps Street
Collingwood
Victoria 3066
Tel. (3) 417 5361
Fax (3) 419 7154

JAPAN

Kinokuniya Company Ltd

17-7 Shinjuku 3- Chome
Shinjuku-ku
Tokyo 160-91
Tel. (03) 3439-0121

Journal Department

PO Box 55 Chitose
Tokyo 156
Tel. (03) 3439-0124

SOUTH-EAST ASIA

Legal Library Services Ltd ·

STK Agency
Robinson Road
PO Box 1817
Singapore 9036

AUTRE PAYS
OTHER COUNTRIES
ANDERE LÄNDER

Office des publications officielles
des Communautés européennes

2, rue Mercier
L-2985 Luxembourg
Tél. 499 28 -1
Télex PUBOF LU 1324 b
Fax 48 85 73/48 68 17

3/93